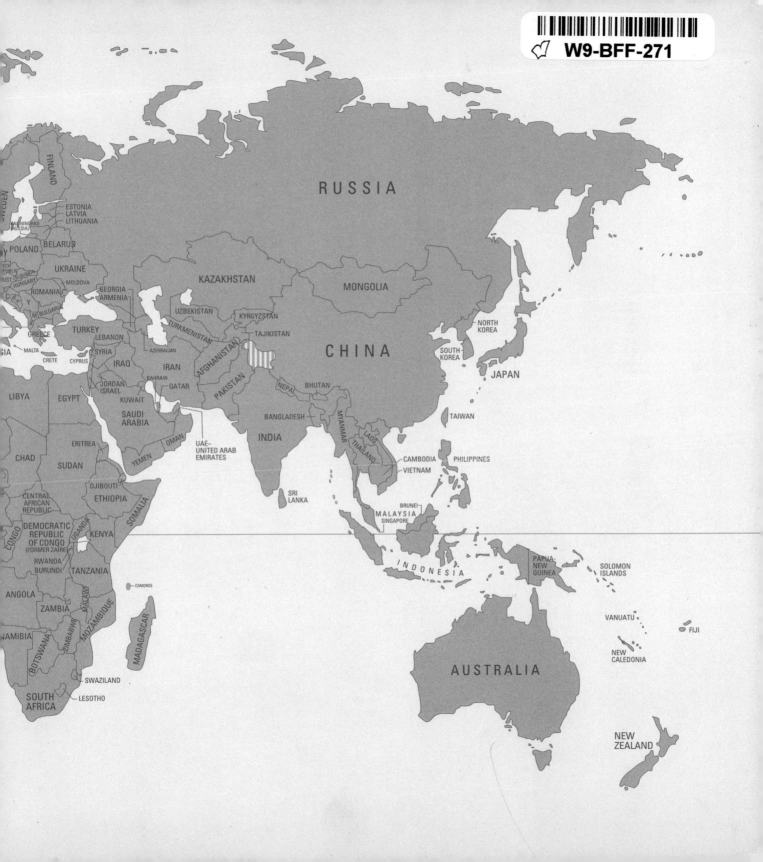

Discovering
the Human World

Discovering
the Human World

Christine Hannell
Stewart Dunlop

OXFORD
UNIVERSITY PRESS

OXFORD
UNIVERSITY PRESS

70 Wynford Drive, Don Mills, Ontario M3C 1J9

Oxford University Press is a department of
the University of Oxford.

It furthers the University's objective of excellence in research,
scholarship, and education by publishing worldwide in

Oxford   New York

*Athens   Auckland   Bangkok   Bogotá   Buenos Aires   Calcutta
Cape Town   Chennai   Dar es Salaam   Delhi   Florence   Hong Kong   Istanbul
Karachi   Kuala Lumpur   Madrid   Melbourne   Mexico City   Mumbai
Nairobi   Paris   São Paulo   Shanghai   Singapore   Taipei   Tokyo   Toronto
Warsaw*

with associated companies in   *Berlin   Ibadan*

Oxford is a trade mark of Oxford University Press
in the UK and in certain other countries

Published in Canada
by Oxford University Press

Copyright © Oxford University Press Canada 2000

The moral rights of the author have been asserted

Database right Oxford University Press (maker)

First published 2000

Canadian Cataloguing in Publication Data
Hannell, Christine
Discovering the human world
Includes index.
ISBN 0-19-541344-X
1. Human geography—Juvenile literature.
I. Dunlop, Stewart. II. Title.
GF43.H36 2000   304.2   C00-930558-0

Printed and bound in Canada
This book is printed on permanent (acid-free) paper ∞

2 3 4 5—04 03 02 01 00

Design: Tearney McMurtry
Illustrations: Ignition Design and Communications
          (Maps) Visutronx Services
          (Technical) Ibex Graphic Communications Inc.

The authors would like to acknowledge those who
made contributions and suggestions in the writing of
this book. Christine Hannell would like to thank:
Anna Hillen, Tez Darnell, Tiina Randoja, Amy Britten
at Guelph Business Development, Denis Roy at
Kirkland and District Community Business Centre,
Jodi Hodginson of the Greater Toronto Airports
Authority, Denise Burlingame and Gloria Irani at Ekati
Diamond Mine, Dofasco, McNeill Consumer Products
Co., University of Guelph Library, Wellington County
Library.

Stewart Dunlop would like to thank: Eric Kramers
of the Canada Centre for Remote Sensing, and
Jennifer Elder.

The publisher would like to thank Rob Mewhinney,
Curriculum Advisor for Social Studies and Geography,
Toronto District School Board; and Theresa Varney,
Lockview Elementary School, St. Catharines, Ontario,
for their comments on the manuscript.

To Chelsea, Quinlan and China

and

To the children of
Kusi village in Eastern Ghana

List of Geo-Tools

List of Case Studies

List of Geography Skills

Contents

Unit One

Discovering Human Patterns

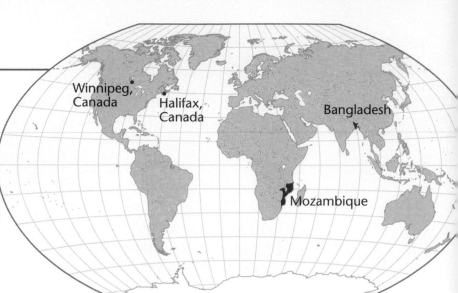

Just as geographers study patterns in the physical world, they also look for patterns in where and how people live. They are interested in why settlements vary in size and other features. These settlements make varied patterns on the land, which geographers want to describe and explain. Geographers are also interested in finding out why people choose to live in particular areas. For example, why are some sections of large cities very crowded, while other parts of the world have no people living in them?

This unit will help you discover the patterns of the world's population. You will see where settlements are located and what factors affect their location. You will learn how people in different cities and countries use land and organize their lives. You will also study patterns in the growth of the world's population. While you do so, you will "visit" places and regions around the world, including those highlighted in the world map on this page.

Settlement Patterns

Discover patterns in the varied settlements in which people live.

Population Patterns

Discover how people are distributed around the world.

The Growth of Cities

Discover the importance of site and situation in the growth of cities.

Land Use Patterns

Discover patterns in land use and social organization in cities.

Population Growth

Discover how the world's population is growing.

Population Contrasts

Discover the contrasts in standards of living between the developed
and developing regions of the world.

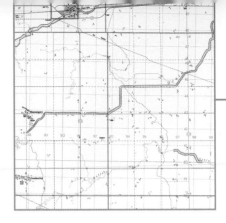

Chapter 1

Settlement Patterns

In this chapter we focus on patterns in human settlements. The information and activities will help you

▸ identify three patterns of settlements
▸ describe how services are affected by settlement patterns
▸ show your understanding of how history and environment affect settlement.

The Origin of Settlements

A *settlement* is any place where people live, or *settle*. A settlement may vary in size from an isolated cottage to the largest city on Earth. In Canada, settlements of over 10 000 people are often referred to as *cities*. Smaller clusters of people are usually called *towns*. Some countries define a town specifically in terms of its population. For example, in Canada, a place may be called a town when it has more than 1000 people. Places with fewer than 1000 people are often referred to as *villages* or *hamlets*.

Settlements exist because people are social creatures. We like company. In the early stages of the human race, people lived in groups of extended family units. They hunted and gathered to survive. In these early days, the land could support few inhabitants. This changed when people learned the skills of farming. Farming made it possible to produce more food than a family needed. Since finding food no longer took up everybody's time, people could turn to other occupations. They could become traders, craftspeople, priests and other types of workers. Surplus farm food was sold to these people, who now clustered together in larger settlements than ever before.

The first sites chosen for settlement were places where people were able to meet their basic needs. The best locations were on dry ground (above flood levels), with access to *water supply*, *fertile soil* (for crops), *timber* (for fuel), and *pastures* (for animals to graze on).

Settlement Patterns in Quebec

People have always settled at sites that can best meet their basic needs. In Canada, during the 17th and 18th centuries, immigrants from France settled on both sides of the St. Lawrence River, where the land was fertile. The river was a vital water supply and transport route for goods coming from Europe. To give all settlers access to the river, the landlords (or *seigneurs*) divided the land into long narrow strips, or lots, at right angles to the waterfront (see the air photograph in Figure 2(b) on page 7).

Figure 1
This location was once part of the ancient civilization of Mesopotamia. Some of the world's first permanent farming villages were set up here. What features in this scene make the area a good settlement site?

This type of property arrangement was called the *seigneurial* system. Each set of long narrow lots was called a *rang*. When French rule ended in 1763, 90 per cent of the population lived within 1 km of the river. Eventually, roads were built through the lots. The roads were built at right angles to the waterfront. Later on, roads parallel to the river were also constructed further inland, and people eventually settled along them.

Although the river is not now as important as it once was, the pattern of settlement remains similar (see Figure 2 on page 7). The seigneurial pattern led to the development of the **linear settlement**.

linear settlement—a pattern of settlement in which homes and other buildings follow the lines taken by roads.

Topographic Maps

Topographic maps show certain features of the Earth's surface in detail. These maps are useful for studying human patterns because they show actual settlements, as well as other features. These kinds of maps allow us to study the physical features of the land on which settlements are built, including the height and slope of the land as well as drainage features such as rivers and marshes. These features are part of the "site" of any settlement.

If the topographic map covers a large enough territory, it also enables us to learn about the "situation" of a settlement—that is its relation to other settlements and to the surrounding land.

Topographic maps also show *contours*—lines that join places of the same height above sea level. The topographic maps in this chapter have been reduced from their original 1:50 000 scale. For the maps on pages 7, 9, 10, and 14, note that each square of the blue grid on these maps shows 1 km². This means that the distance between each pair of blue lines represents one kilometre on the ground.

WEB LINK

To learn more about topographic maps, look up http://maps.NRCan.gc.ca/maps101/index.html To see Canadian topographic maps on two different scales, look up http://maps.NRCan.gc.ca/topographic.html

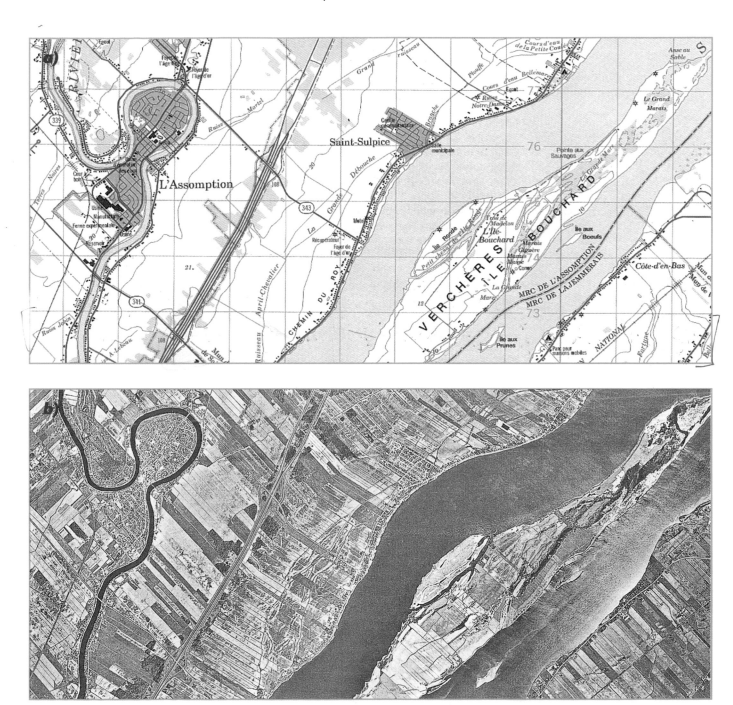

Figure 2
The topographical map in (a) and the air photograph (b) show the seigneurial pattern of settlement east of Montreal.

Discover For Yourself

1. Explain how the seigneurial system led to a linear pattern of settlement in Quebec.
2. What disadvantages might the seigneurial system have? Explain your answer.
3. Sketch a map of the part of Quebec shown in Figure 2 on page 7. Your map should show
   a) the land and the river
   b) the road pattern
   c) the linear settlements
   d) the shape of the fields.

Settlement Patterns in Ontario

In 1760, when British immigrants began to settle in Ontario, it was called "Upper Canada." The British were joined by Empire Loyalists from the United States in the early 1780s. In what is now known as the province of Ontario, most of the land was divided up into rectangular townships parallel to the shores of the lakes. These townships were further divided into farms, separated by roads. The government decided to intersperse these properties with lands reserved for the clergy and for the crown. This created a checkerboard pattern, with farming settlements separated by reserved lands.

However, the government decided additional roads were needed to transport troops, which would allow Upper Canada to defend itself against attack. These roads allowed farming families to travel and settle on farmlands further along Lake Ontario and Lake Erie.

As a result, the rural population pattern in Figure 3 is more scattered than in Quebec—the symbols showing settlement are more widely spaced apart. But, it is also partly linear, as farms are found along the roads that are parallel to the lakeshores. Rural settlement in much of Ontario today remains partly linear and partly scattered.

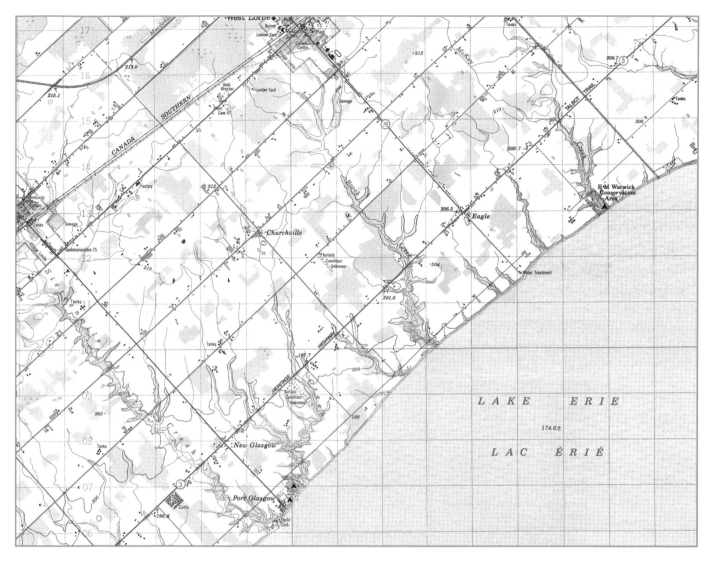

Figure 3
Rural Settlement Pattern on the
North Shore of Lake Erie

Settlement Patterns in Atlantic Canada

Other regions of Canada also have distinct patterns of settlement.
Their physical geography is a major factor in these patterns. For
example, Newfoundland has a rocky landscape, with only small
patches of good soil. Fishing was the dominant industry from
the early days of settlement until the late 20th century, when the
fishing industry declined drastically. Throughout this time, most

Newfoundlanders lived in tiny fishing hamlets, strung along the numerous bays and inlets of the rugged coastline. In the 1960s, the government of Newfoundland brought in a plan to consolidate many of these tiny settlements into larger centres. This made it easier to organize and operate public services. For example, rather than each hamlet being responsible for its own school, schools could be run from a central office for a larger area.

In spite of these developments, Newfoundland's population remains largely coastal, as Figure 4 shows.

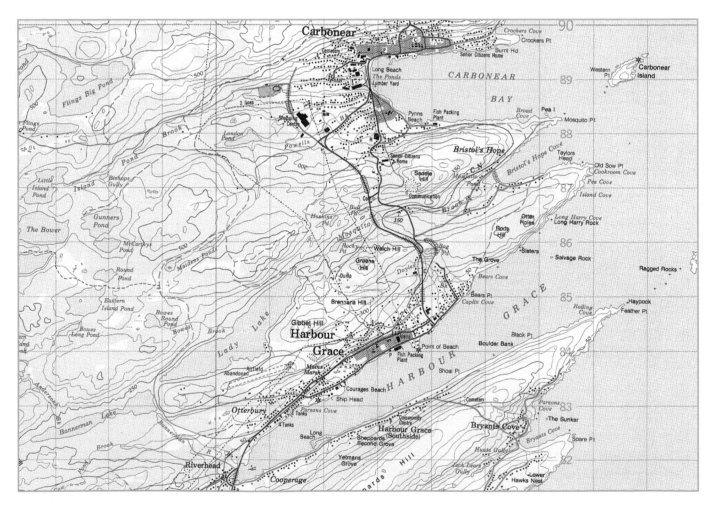

Figure 4
Many of the fish packing plants shown in this map have already been closed.

Discover  For Yourself

1. Form small groups. In your groups, make a copy of the following
   organizer. Decide on the answers to each section and fill in your
   organizer. Compare your group's organizer with those of other
   groups. Add any information your group might have missed.

Settlement Pattern	Physical Features	Shape of Settlements	Size of Settlement Clusters
Southern Ontario			
Newfoundland			

2. Discuss these questions in your groups. Give reasons for the
   answers you give.
   a) Have historical events played a large part in the Ontario
      pattern?
   b) Have historical events played a large part in the
      Newfoundland pattern?

Settlement Patterns in Western Canada

The physical geography of Canada's western regions is varied. On the one hand, the Prairie provinces are very flat. In contrast, only a small proportion of British Columbia consists of low-lying, flat land. Most of such land in British Columbia lies in the lower Fraser Valley, where population growth has taken place rapidly. The rest of the province is largely mountainous, so settlements are located in valleys. These settlements are mainly related to the resource industries of British Columbia—particularly forestry and mining. On the coast there are some fishing and logging towns.

Case Study   *Settlement Patterns on the Prairies*

With no mountains and few major rivers, the Prairie provinces are much flatter than many other parts of Canada. Such a landscape allows settlements to be arranged in a very regular pattern. The *Homestead Act* of 1872 divided Western Canada into townships, each consisting of 36 square-mile-sections (1.6 km by 1.6 km). Each section was divided into four quarter-sections of 65 ha. The quarter-sections, called *homesteads*, were sold to people who wanted to become farmers. The price was only $10. Land was also given to railway companies to encourage the building of railway lines across the Prairies. The pattern of the land when it had been divided up looked like a large checkerboard, with homesteads alternating with railway land.

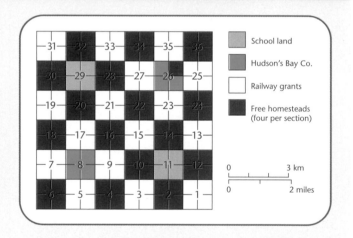

Figure 5
This is the plan of a township in Western Canada. Note that small amounts of land were reserved for the Hudson's Bay Company and for schools.

O━ᴍ   **scattered settlement**—a pattern of settlement in which houses and other buildings are placed a long distance apart from each other.

clustered settlement—a pattern of settlement in which houses and other buildings are laid out closely together.

You can see this pattern today if you fly over the Prairies.

Farming settlement on the Prairies may be described as **scattered**. This is because the homestead on which each farming family built their house covered a fairly large area (a quarter of a square mile). You can see many such scattered farms in Figure 7 on page 14.

Figure 7 also shows some **clustered settlements**. Plum Coulee and Horndean are examples of villages that grew up along the railway to provide services to farmers. Gnadenthal is one of several villages in southern Manitoba that were founded by Mennonites who came from Eastern Europe in the late 19th and early 20th centuries. Mennonite farmers were given blocks of land rather than quarter-section homesteads. They preferred to build villages consisting of side-by-side homes. Each farmer walked to their fields. Blumengart is a colony of Hutterites—another immigrant group from Eastern Europe—who lived communally rather than adjacent to one another.

Roads, Railways and Settlement

In most parts of the world, including eastern Canada, *subsistence farming* was the first type of farming practised. This means that farmers produced food for themselves and their families rather than for sale. On the Prairies, however, farming was *commercial* from the start. By the 1890s, prairie farmers were producing wheat to sell to the rapidly expanding markets of Europe. To do so, they needed a link with the rest of the world. The railways provided that much needed link.

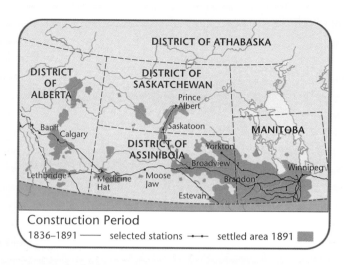

Figure 6
Railway Construction on the Prairies, 1836–1891

The railways influenced the pattern of settlement on the prairies. The branch railway lines could not be more than a half-day journey (by horse and cart) away from farms. This allowed farmers to deliver their grain to storage "elevators" and get back home on the same day. As a result, railway lines were built no more than 20 miles (32 km) apart. Clustered settlements, like those in Figure 7, grew up at regular intervals along the railway lines.

Eventually, horses and carts were replaced by trucks, and roads were improved. Farmers with trucks could deliver their grain much more easily to the elevators. This led to the closing of many branch railway lines. As a result, many small settlements on the prairies have disappeared, and the people have moved to larger centres.

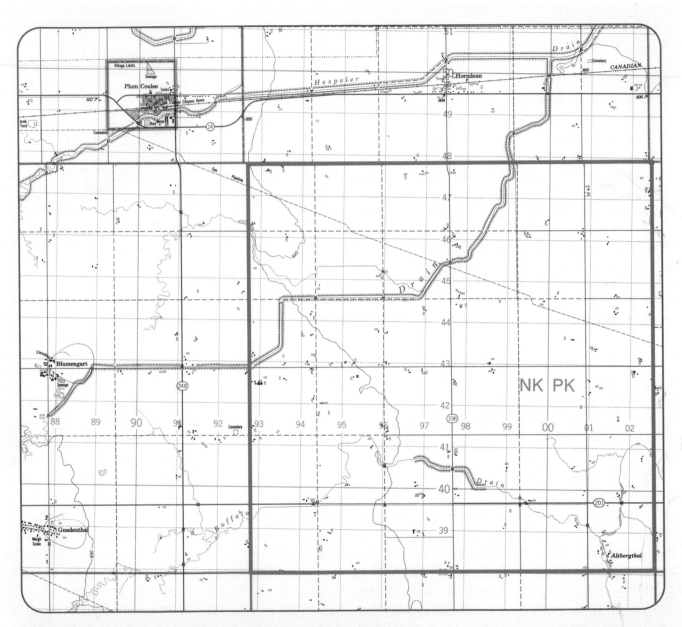

Figure 7
This is a topographic map of part of southern Manitoba. Note that on topographic maps in Western Canada there are two separate grids: the square-kilometre grid (blue lines) and the square-mile grid used for townships and sections.

 With Maps

1. Form small groups. In your groups, discuss and answer these questions about Figure 7.
   a) What evidence is there that this part of Manitoba is flat?
   b) Calculate the approximate number of people who live in the township that is outlined. Assume that each tiny cluster of black squares is a farm with four people.
   c) Why do you think there is not much scattered settlement close to the Mennonite colony of Gnadenthal or the Hutterite colony of Blumengart?
   d) All roads (red and orange lines) follow the north/south and east/west boundaries of sections. Do you think this is an ideal pattern for roads in rural areas? Give reasons for your answer.

Turn to page 258 to learn how contours are used to show land height on topographic maps.

Settlements and Services

All settlements provide services for the people who live in and near them. Some types of services are found in even the smallest settlements (villages or hamlets). Other types are only found in larger settlements. Figure 8 on page 16 gives examples of some patterns involving different types of stores.

Why do you think grocery stores are found in all sizes of settlements while furniture stores are not? One reason is that people buy food often and therefore like to live near a source of food. On the other hand, people may buy furniture only once every few years. They are willing to travel to find what they are looking for.

Grocery and other stores we use frequently are called *low order services*. Another example of a low order service is a gas station. People need these services on a regular basis, and they do not need to travel far to reach them. Furniture stores and car dealerships are *high order services*. Somewhere between high order and low order services are *middle order services*, such as shoe and clothing stores.

Type of store	Minimum size of settlement	Distance to travel	Possible frequency of use
Grocery store	Village or hamlet	Walking distance	At least once per week
Shoe store	Small town	Short ride	Once or twice per year
Furniture store	Mid-size town	Longer ride	Once every year or two
Superstore	Large town	Longer ride	Several times per year

Figure 8
Think of three other types of stores to add to this organizer. How would you fill in their patterns?

All these services are *retail services*—shops selling goods to the public. Superstores, for example, sell a wide range of products, often at discounted prices. However, they are only found in larger urban centres. *Wholesale services* are provided in larger centres as well. Wholesalers are distributors who buy from the manufacturers and sell goods to the shops.

Other services besides retailing that settlements can provide include entertainment (for example, cinemas) and financial services (for example, banks).

Many residents are employed in services while others are employed in industry. In cities such as Hamilton, a large number of inhabitants earn their living in industries associated with steelmaking. Other towns may be ports, or holiday resorts (for example, Acapulco in Mexico).

The larger a settlement is, the greater the number and type of services it provides. Figure 10 shows how this is true for the 10 *census metropolitan areas* (CMAs) in Ontario. CMAs include central cities together with surrounding suburbs.

Figure 9
Hamilton is an industrial city.

Census Metropolitan Area (CMA)	Population 1996	Total Number of Stores	Grocery and Supermarket	Women's Clothing	Shoe Stores
Toronto	4 263 757	5 725	611	721	379
Ottawa–Hull	1 010 498	1 591	132	234	101
Hamilton	624 360	1 067	122	126	66
London	398 616	853	121	113	57
Kitchener	382 940	640	121	67	29
St. Catharines–Niagara Falls	372 406	596	153	40	19
Windsor	278 685	423	90	40	20
Oshawa	268 773	322	67	25	14
Sudbury	160 488	241	16	33	11
Thunder Bay	125 562	194	44	18	6

Figure 10
The CMAs of Ontario

  **For Yourself**

1. Why is the average distance travelled to a grocery store less than to a furniture store?
2. The CMAs in Figure 10 are ranked from largest to smallest populations.
   a) Rank the CMAs from largest to smallest total number of stores.
   b) How does your ranking in (a) compare to the ranking in Figure 10?
   c) What does the comparison tell you about the relationship between population and number of stores?

Turn to page 97 to discover another way of showing this relationship.

Summary

In this chapter you have discovered that settlements first came into being in places where the basic needs of people were met. You have studied the many patterns in which settlements can be arranged on land. You have also learned that settlements provide services for the people who live in and near them.

Reviewing Your Discoveries

1. Summarize the reasons why some rural settlement is clustered and some is scattered.
2. Describe how the closure of a branch railway line on the Prairies can affect farmers who live close by.
3. Explain the advantages of locating middle and high order services in larger towns and cities.
4. Explain how improving the road system might affect the services provided by
   a) small hamlets
   b) larger towns

Using Your Discoveries

1. Conduct a survey to find patterns in people's use of stores at a local shopping centre. Follow these steps:
   a) Working in groups, construct a questionnaire response sheet like the one in Figure 11. Label the three rightmost columns with a low order service (such as a grocery store), a middle order service (such as a shoe store) and a high order service (such as a furniture store). Choose services that are found at a local shopping mall.
   b) Visit the shopping mall and ask 30 people close to each of the three services these two questions:
      – How often do you use the service?
      – How far have you travelled to use the service?
      Record each answer with a checkmark in the appropriate space of the response sheet.
2. In your groups, show the results of your questionnaire with six bar graphs. Follow these steps:
   a) Draw three bar graphs to show "Frequency" results (one bar graph for each store). Use the type of store as the title of each bar graph. On the horizontal axis, mark the three categories of frequency. On the vertical axis, mark the number of responses from 0 to 30. Then draw bars to show the number of people answering yes to each frequency category for the store. Repeat for the two other stores.

		Grocery store	Shoe store	Furniture store
Frequency	Once or more per week			
	Once per month			
	Once per year or less			
Distance travelled	Less than 1 km			
	1 – 5 km			
	Over 5 km			

Figure 11
Use a survey chart to keep track of peoples' responses.

b) Draw another three bar graphs to show "Distance travelled" results. Again, use the type of store as the title of each bar graph. On the horizontal axis, mark the three categories of distance. On the vertical axis, mark the number of responses from 0 to 30. Then draw bars to show the number of people answering yes to each distance category for the store. Repeat for the other two stores. A model of one of these graphs is shown in Figure 12.

3. Compare the two graphs for each store. On your own, write one page summarizing your findings.

4. Based on your experience, would you change the questionnaire in the future? If so, how?

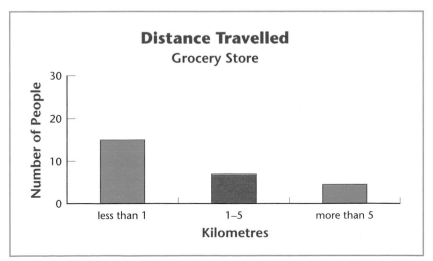

Figure 12
Use this graph as a model for your answers to Question 2 a) and b).

Chapter 2

Population Patterns

Key terms

O—π

census
population density
Industrial Revolution

O—π

census—a count of all the
people in a country, together
with some details about them.

In this chapter we focus on population patterns in Ontario, Canada and the world. The information and activities will help you

▸ identify and describe the characteristics of places with high and low population densities
▸ show your understanding of how history affects population distribution
▸ show your understanding of how environment affects population distribution.

Counting People

Almost every country in the world conducts a population **census**. Do you remember the last time a census took place in Canada? Canada has a population census every five years. The last one in the 20th century took place in 1996.

The information recorded by Statistics Canada in the Census of Population can be divided into two categories:

Things about a person that cannot change. Examples:	Things about a person that may change. Examples:
– place of birth	– place of residence
– date of birth	– marriage status
– ethnic origin	– number of children
	– job
	– income
	– items owned (e.g., car, computer)

The information gathered in the census helps the government understand the state of the economy. It also helps the government measure or determine social conditions, such as poverty levels and crime rates. Social services agencies and police use the same data to help them solve social problems.

To get its population data, Statistics Canada first counts people in each *household* across the country. Household counts are then grouped into *enumeration areas* (*EAs*). Each EA usually consists of about 300 to 400 households. In rural areas, EAs are combined into larger units, called *census subdivisions*. Census subdivisions are then grouped into even larger *census divisions*. In cities (called *metropolitan areas* in the Census), EAs are grouped into *census tracts*. Figure 2 shows how this is done.

Figure 1
Census data can be used by governments to plan for new hospitals and schools, and by companies to locate new factories or services.

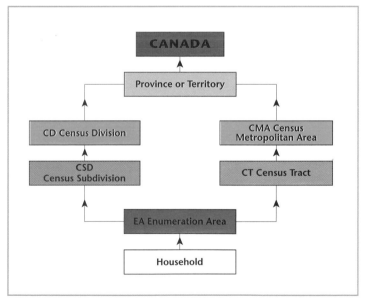

Figure 2
This diagram shows how census data for households is combined to provide data for the provinces/territories and Canada.

WEB LINK

To find out more about the Census of Canada, look up http://www.statcan.ca/english/census96/list.htm

Discover With Tables

1. Figure 3 is a table of Canada's census metropolitan areas (CMAs) from the 1996 Census.
   a) Which city has shown the greatest increase in total population since the 1991 Census?

Census Metropolitan Area (CMA)	Population 1991 (000s)	Population 1996 (000s)	Percent change 1991–1996
1. Toronto	3 899	4 264	9.4
2. Montreal	3 209	3 327	3.7
3. Vancouver	1 603	1 832	14.3
4. Ottawa–Hull	942	1 010	7.3
5. Edmonton	841	863	2.6
6. Calgary	754	822	9.0
7. Quebec	646	672	4.1
8. Winnipeg	660	667	1.0
9. Hamilton	600	624	4.1
10. London	382	399	4.5
11. Kitchener	356	383	7.4
12. St. Catharines-Niagara	365	372	2.2
13. Halifax	321	333	3.7
14. Victoria	288	304	5.7
15. Windsor	262	279	6.3
16. Oshawa	240	269	11.9
17. Saskatoon	211	219	3.8
18. Regina	192	194	1.0
19. St John's	172	174	1.3
20. Sudbury	158	160	1.8
21. Chicoutimi-Jonquière	161	160	–0.3
22. Sherbrooke	141	147	4.7
23. Trois–Rivières	136	140	2.7
24. Saint John	126	126	–0.1
25. Thunder Bay	125	126	0.5

Figure 3
Canada's Census Metropolitan Areas Ranked by 1996 Population

b) Which city has shown the greatest percentage increase since the 1991 Census? Explain why the answers to a) and b) are different.

2. Is it generally true that the larger CMAs in 1991 experienced the greatest growth from 1991 to 1996? Give reasons why larger cities may grow faster than smaller cities.

3. Canada's resource industries suffered some economic decline in the 1990s. Certain cities depend on these resource industries. When the industries suffer, the population of the cities may show slower growth or even a decrease.
   a) Which city or cities do you think might have been affected by the weak market for wheat?
   b) Which city or cities do you think might have been affected by the weak market for metal ores?

Population Density

We saw in Chapter 1 that the population in farming areas is mainly scattered, with only small clusters of people in service centres. Cities, on the other hand, have many people living in a relatively small area. The term that describes this difference between *rural* areas and *urban* areas is population density. Population density is usually calculated by dividing an area's total population by the number of square kilometres in the area.

Follow these steps to work out the population density within your classroom:
1. Measure the length and width of the room to the nearest metre.
2. Multiply the length by the width to get the area of the room in square metres.
3. Count the number of people in the room.
4. Divide the number of people by the area in square metres. (This figure is the *density per square metre*.)
5. Multiply by 1 000 000 to get the *density per square kilometre* (since there are 1 million m² in 1 km²).

Figure 4
Hong Kong has one of the highest population densities in the world.

population density—a
measure of how many people live
in a unit of area, usually a square
kilometre.

You may be surprised at how high the **population density** in your classroom is. The density is always lower when we calculate it for areas in cities and towns, rather than for rooms where people are crowded together.

Discover *With Maps and Tables*

1. Calculate the population density in the township outlined on Figure 7 on the page 14. Each section is one square mile, or approximately 2.5 km² in area. Assume four people per farm and that each small cluster of black squares is a single farm.
2. Divide into groups of four to fill in the organizer in Figure 5. Follow these steps:
    a) Make a group copy of the organizer. Each member of the group should choose to work with one of the following topographic maps in Chapter 1 (make sure every person chooses a different map): Figure 2 on page 7, Figure 3 on page 9, Figure 4 on page 10 or Figure 7 on page 14.
    b) On your map, choose 10 one-kilometre squares (outlined by the blue lines). Make sure that you include a variety of different types of settlements, including any squares with no settlement.
    c) Ignoring any towns that may be in the squares, estimate the number of houses in each square.
    d) Assume four people per house. Multiply the number of houses by four to get the total population in each square. Record these population figures in your row of the organizer.
    e) Add up the total population of all 10 squares. Record the sum in the organizer.
    f) Divide the total by 10 to get the average rural population density per square kilometre. Record your answer in the organizer.
    g) As a group, list the factors that help to explain the differences in density between the four areas.
3. Divide into small groups. Each group should choose to work with Canada as a whole, or a different province or territory.
    a) In your groups, use the data in Figure 6 (on page 25) to calculate the population density for the area you have chosen.

MATH
LINK

Map extract	Square										Total	Density/ km²
	1	2	3	4	5	6	7	8	9	10		
Quebec (Figure 2 on page 7)												
Ontario (Figure 3 on page 9)												
Newfoundland (Figure 4 on page 10)												
Manitoba (Figure 7 on page 14)												

Figure 5
Organizer for Question 2 on Page 24

b) Compare the population densities to identify your region as high or low density (Canada as a whole has a low population density). Use an atlas to find reasons for the high or low density of your region.

Province/Territory	Area (000 sq. km)	Population (000s) in 1996
Newfoundland	406	552
Prince Edward Island	6	135
Nova Scotia	55	909
New Brunswick	73	738
Quebec	1 541	7 139
Ontario	1 069	10 754
Manitoba	650	1 114
Saskatchewan	652	990
Alberta	661	2 697
British Columbia	948	3 724
Yukon Territory	484	31
Northwest Territories	3 426	64
Canada	9 971	28 847

Figure 6
This chart gives the area and population data for Canada and its provinces and territories in 1996. Nunavut only came into existence in 1999, so it could not be included in this chart.

Case Study   *Population Density in Ontario*

If you have ever travelled around the province of Ontario, you have probably seen many types of contrasts between regions. Some of these contrasts involve population density. For example, what can you predict about the population density of the three Ontario places shown in Figure 7? Figure 8 shows the different densities in the Census Subdivisions of Ontario. Each colour shade represents a different range of density.

The highest density is found in Toronto's York, with 6279 people per km². Activities in this high-density region range from industry and transportation to finance, business and education. By far the largest number of stores in Ontario are located in Toronto.

Figure 7
Population densities vary greatly between vast urban centres, southern farming regions, and cold northern areas of Ontario.

Compare these conditions of city life with the life of a farmer in southern Ontario. He or she lives in a rural area, such as the one shown in Figure 7 b). There are fewer people in the area

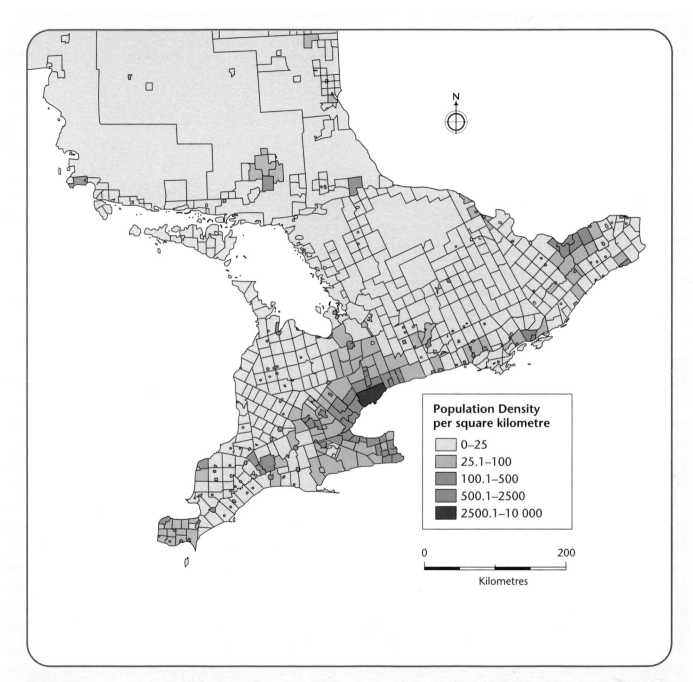

Figure 8
Population in Ontario by Census Subdivision, 1996

Population Density per square kilometre
- 0–25
- 25.1–100
- 100.1–500
- 500.1–2500
- 2500.1–10 000

as compared to the city. But most people know each other well and there is a strong sense of community.

The lowest population density in Ontario is in Rainy River (northwest Ontario): only 0.1 people per km². Population here and throughout north and northwest Ontario is scattered, with large areas that have no inhabitants at all. Most people live in mining settlements or in small service centres along roads.

Discover For Yourself

1. Look back at the average population density for the whole province of Ontario that you calculated for Question 3 on page 24. Why is this a misleading statistic? (Hint: consider where most of the people of Ontario live).
2. Make a list of reasons why most people in Ontario live within a short distance of the Great Lakes.
3. Why do you think the population in Northern Ontario is scattered?

Population Distribution

The list that you made for Question 2 above can help you understand Ontario's *population distribution*—where people in Ontario live. Population distribution can be shown with coloured regions, as in Figure 8, or with dots, as in Figure 9. A striking pattern shown in Figure 9 is that most Canadians live within 300 km of the United States border.

Population distribution throughout the world follows patterns that are the result of historical events. Some of the most important of them occurred in the late 18th century. At this time, farmers in England were allowed to enclose their fields. For the first time, farmers worked their own individual farms, rather than cultivating common fields. With this change, new crops, such as turnips, were introduced. *Crop rotation* (changing the types of crops grown in each field from year to year) became widely practised. This helped to preserve soil fertility and prevent crop disease. Farming became more productive and could support more people.

One red dot represents 1000 persons

One black dot represents 100 persons north of latitude 60°N

● cities with more than 20 000 inhabitants

settled area

0   500   1000
kilometres

Figure 9
This map shows the actual distribution of population in Canada, with each dot representing 1000 people (except in the Far North, where each dot represents 100 people).

Industrial Revolution—a set of changes in technology, social life, and politics that occurred during the late 18th and early 19th centuries. During this time coal was used to power the steam engine and many other mechanical inventions. As a result, Europe, particularly Great Britain, became the leading industrial region of the world.

Shortly after these advances in agriculture occurred, the **Industrial Revolution** began. This historic event created the demand for labour in towns. The more productive farms provided food for these growing industrial centres. While the population in rural areas grew rapidly, machinery began replacing many farm workers, who then had to migrate to cities to find work. As the Industrial Revolution advanced, population in Western Europe surged. By the 20th century, Western Europe had a dense population, supported by industry and commercial farming.

In parts of the world where more traditional types of farming are still practised, dense populations have also developed. This is especially true in India (where more than half the population are farmers) and China (where nearly half the population are farmers). Farming is less mechanized in these countries than in Europe and North America, and more use is made of human and animal labour. Farming is mainly *subsistence*, rather than *commercial* (that is, families grow food mainly for their own needs, rather than for sale). Figure 11 shows the large, dense populations of India and China, which together make up over one third of the world's people.

Discover *With Photographs and Maps*

1. Divide into small groups and make a copy of the organizer in Figure 10. In your groups, fill in the organizer, using the two photographs in Figure 12 on page 32 and the map in Figure 11.

Figure 10
Organizer for Question 1

	India	Canada and the United States
Size of farm		
Type of farming practised (subsistence or commercial)		
Farming tools used		
Destination of farm products		
Income level of farmer		
Population density of country		

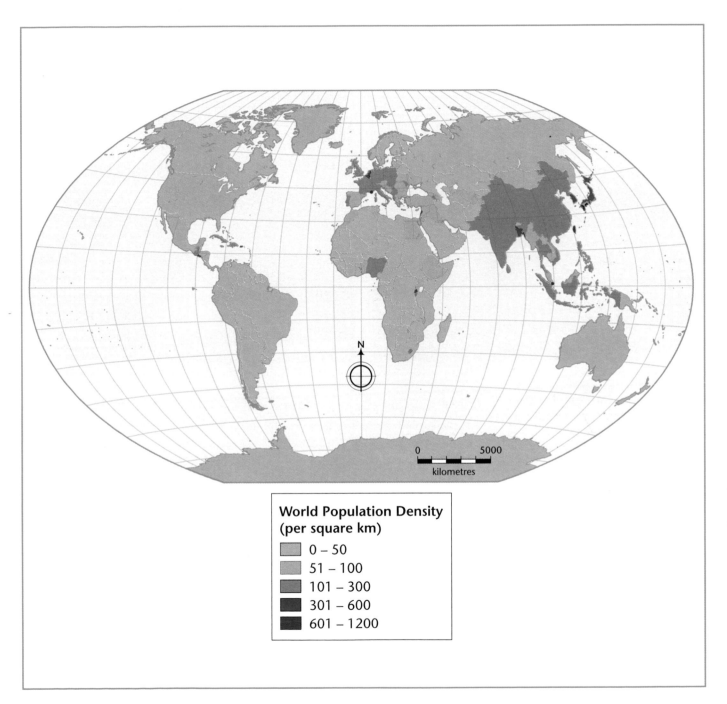

**World Population Density
(per square km)**

- 0 – 50
- 51 – 100
- 101 – 300
- 301 – 600
- 601 – 1200

Figure 11
Population Densities of Countries Around the World

Figure 12
Farmers in India are more reliant on animal labour than farmers in developed countries such as Canada, where machines and tractors are common.

2. In your groups, brainstorm what you know about farming. Use your ideas to explain why the rural population density in North America is much lower than in India and China.

The Environment and World Population

If you had the choice to live in any environment, which would you choose? Cold or warm? Dry or moderately wet? Rugged and mountainous or gently rolling in landscape? Your answers probably help explain this general pattern: relatively few people live in areas with a difficult physical environment. There are specific features that make an environment difficult. The environment can be:

▸ too dry
▸ too cold
▸ too high or steeply sloping in its landscape
▸ unsuitable for crop growing because of soil problems

A climate is *too dry* when precipitation is less than the amount of water that evaporates. Soils dry out in this environment, and crop growing is difficult or impossible without irrigation. The only other water supply in dry environments, such as the world's deserts, is from occasional small oases (where water in the ground rises to the surface).

If the environment's climate is *too cold*, people cannot live by farming. Most crops need a temperature of above 6°C in order to grow. Therefore, in much of northern Canada, the growing season is not long enough or warm enough for crop growing. People can live in such areas, but food has to be brought in from farming areas in the south. Without food easily available, not many kinds of communities can develop. Typically, the ones that do are small and scattered.

Since *mountainous areas* are at high elevations, they are likely to be too cold for crop growing. In addition, crop growing is difficult on steeply sloping land because soil erodes easily when heavy rain falls. It is also difficult and dangerous to operate machinery in steeply sloping regions.

There are a few areas of the world where temperature and rainfall conditions are suitable for crop growing, but *soil problems* prevent the success of farming. The best example is the tropical rain forest. In locations such as the Amazon Basin, soils are quickly eroded when exposed to intense sunshine and heavy rain. Farming in this region is possible only when these poor soils get a steady supply of nutrients from leaves falling from rain forest trees. When these trees are cut down, tropical soils can become sterile and unproductive.

When we divide up the world's land into difficult environments (those described above) and favourable environments, we get a ratio of about 80:20. This means that 80 per cent of the world's land poses challenges for human life. And, following the pattern presented at the beginning of this section, only about 10 per cent of the world's population can be found on this land, in thinly scattered settlements. That means that about 90 per cent of people live at a high density on the remaining 20 per cent of land!

Figure 13
This desert land in Sudan could not be used to grow crops without an irrigation system.

Discover With Graphs

1. Use the climate data in Figure 14 to draw climate graphs for Churchill (Manitoba) and In Salah (Algeria). When completed, draw a horizontal line across each graph at 6°C to indicate the minimum temperature for crop growth.

Churchill	Jan.	Feb.	Mar.	Apr.	May	Jun.	Jul.	Aug.	Sep.	Oct.	Nov.	Dec.	Year
Temperature (°C)	−27.4	−26.2	−20.3	−10.1	−1.4	6.0	12.1	11.4	5.6	−1.7	−12.9	−22.6	−7.2
Precipitation (mm)	14.8	12.1	18.2	23.1	27.3	43.0	54.6	61.7	53.3	43.7	31.4	18.3	401.5

In Salah	Jan.	Feb.	Mar.	Apr.	May	Jun.	Jul.	Aug.	Sep.	Oct.	Nov.	Dec.	Year
Temperature (°C)	14.3	16.8	20.9	25.2	30.5	35.7	36.5	36.5	33.0	26.8	20.2	14.0	26.1
Precipitation (mm)	1.6	3.4	1.2	2.0	0.4	0.1	0.0	0.3	0.5	1.6	1.2	3.0	15.3

Figure 14
Temperature and Precipitation Data for Churchill and In Salah

2. Discuss these questions for each location above:
   a) Which type of difficult environment exists?
   b) Why can't a large population exist without importing food?
   c) How could people who live in these environments grow some of their own food?
   d) What activities other than growing crops could support the population?

Figure 15
Churchill, Manitoba is located just south of the 59°N parallel.

Summary

In this chapter you have learned about two important population patterns: density and distribution. You have discovered how information about these patterns can be gained from a population census. You have also seen that distribution patterns are closely related to historical patterns of farming and industry. Finally, you have reflected on the relationship between population distribution and physical geography.

Reviewing Your Discoveries

1. Review your answers to Question 3 on page 24. Use the figures you calculated to make a population density map of Canada. Get an outline map of Canada from your teacher and then follow these steps.
   a) Divide the densities of Canada's provinces and territories into four categories: *well above average* density, *above average* density, *below average* density, and *well below average* density.
   b) Assign a different colour shade to each category and then colour in the provinces and territories on your map. Use darker or brighter colours for denser regions and lighter or pale colours for less dense regions.
   c) Add a title and legend to your map.
2. Give an example of how environment affects population density.

Using Your Discoveries

1. Take a census of your class. Follow these steps.
   a) On poster paper that your teacher has placed on the classroom walls, fill in data on yourself in each of these categories:
      – Your age (to the most recent quarter year, for example 13.25 years)
      – Number of people in your household
      – Approximate distance you travel to school (in kilometres to one decimal point, for example 1.3 km)

b) Divide into three groups to make a class profile using the poster paper information. Using Figure 16 as a model:
   - Group 1: make a bar graph showing students' ages and calculate the *average* age of a student in the class.
   - Group 2: make a bar graph showing household populations and calculate the *average* household population.
   - Group 3: make a bar graph showing the distance travelled to school and calculate the *average* distance.

2. Choose a block close to your school. The block should be as rectangular as possible and should not include apartment buildings. Calculate the approximate population density of the block. Follow these steps.

a) Calculate the area of the block by measuring its length and width (in metres) and multiplying them together. (Include half the width of the roads that surround your block in your measurements.) Divide by 10 000 to get the area of the block in hectares (1 ha = 100 m² or 10 000 square metres).

b) Multiply the *average household population* of a student in your class (calculated in Question 1(b)) by the number of houses in the block. This gives you an idea the block's total population.

c) Divide the population by the area in hectares to get the population density per hectare. Finally, multiply by 100 to calculate the density per square kilometre (1 km² = 100 ha).

d) What problems would there be in following the above steps if the block contained apartment buildings?

MATH
LINK

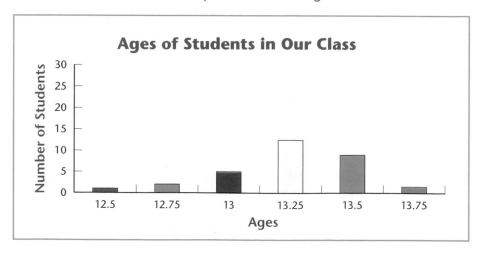

Figure 16
Use this graph as a model for answering Question 1 b).

Chapter 3

The Growth of Cities

In this chapter we focus on the characteristics of cities. The information and activities will help you

- show your understanding of how "site" influences settlement
- show your understanding of how "situation" influences settlement
- show your understanding of factors that affect urbanization.

Where Cities Are Built

Look into the future and imagine that you are going to build a house. You are faced with two important decisions. First, you must decide on a **site** on which to build. The site should be stable, so that it provides a secure foundation for the house. For example, you do not want to build at the edge of a coastal cliff where the sea is eroding the land. Neither do you want to build in an area where there is a high risk of an earthquake.

Second, you must decide on a good relative location, or **situation**, for your house. Examples of good situations are

- close to transport routes
- close to shops
- close to your place of work

Site and situation are important factors in understanding the rapid growth of cities throughout the world. When you read about different cities in this book, try to speculate on the effects site and situation may have had on the location, growth, or lack of growth of each.

Key terms

site
situation
urbanization

site—the ground on which a building or city is built.

situation—the location of a building or city in relation to surrounding places.

Figure 1 (a–d) shows some types of sites on which towns
may grow:

(a) A river crossing point (at a ford, or where a river narrows to
    allow a bridge to be built)

(b) A defensive site (on a hill, or surrounded by the bend of a
    river)

(c) A good natural harbour (for example, the settlements in
    Figure 4 on page 10)

(d) An island in a river (for example, the city of Paris was sited
    on an island in the River Seine).

    Other favourable sites on which cities may develop include:

    ▸ A river bank that is high and dry (to avoid the danger of
       flooding)

    ▸ An area with surrounding flat land (to allow towns to
       expand).

    Cities face problems when they grow on unfavourable sites.
For example, San Francisco was sited in an area without surrounding flat land. When the city grew, buildings had to be built
cramped together without gardens.

    Hamilton's site has had an interesting effect on its develop-
ment. The site is on Burlington Bay, at the west end of Lake
Ontario. It can be divided into a lower part to the north and a
higher part (called the "mountain") to the south. The steep slope
of the Niagara Escarpment runs across the site, dividing the lowland from the mountain.

    This site had many advantages for settlement. One advantage
was that the many streams running down the Niagara Escarpment
helped to run mills and supported early industry in the 19th
century. A second advantage was its situation near Lake Ontario,
making it an important transportation centre.

    Since the lowland area adjoined the waterfront, it was this area
on which heavy industry developed. Canada's largest steel factories
were built here. In the early days of steel production, air pollution
was severe. Winds, which often came from the west, blew the
pollution eastward. The land above the Escarpment escaped much
of this pollution. As a result, this land became a sought-after
residential area. Land on the mountain is more expensive and
homes are higher-priced there than elsewhere in the city.

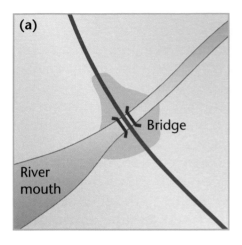

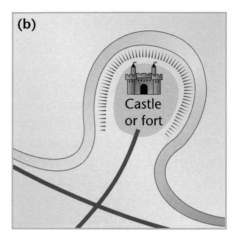

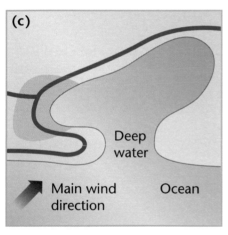

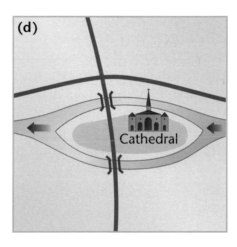

Figure 1
The first diagram (a) shows a river crossing, (b) a defensive site protected by water, (c) a natural harbour, and (d) a river island.

Discover With Maps

1. Divide into small groups. In your groups, look at the following figures of sites and answer these questions.
    a) What was the main site advantage for L'Assomption (Figure 2 on page 7)?
    b) What are the site advantages of the coastal settlements in Newfoundland (Figure 4 on page 10)?
    c) What advantages have settlements on hills had?
    d) What site disadvantages have they had to overcome?

The Situation of Cities

When you look at a political map of Canada, you find that it has only a small number of really large cities. But it has many smaller cities and towns. Most of these smaller places have at least some of the site advantages we have just studied, but have failed to grow into large cities. Why do only a few settlements grow into large cities, while most remain as villages or towns?

One reason is that a region or country does not need many big cities. This is because people do not often use the high-order services that big cities offer. They are willing to travel some distance to reach them. Therefore only a few large cities are needed in a country.

A second reason is that only a few places have the special advantages that help a city to grow. When a city does grow, it limits the growth of places around it. For example, Montreal in the St. Lawrence valley is located where the Ottawa River joins the St. Lawrence River. It is therefore situated at the centre of several transportation routes, including a southern link to New York via the Hudson River. This advantage of a "route centre" situation has helped Montreal grow at the expense of nearby centres.

Figure 2
How often do you think people from a smaller centre would use the services shown here? Would they be willing to travel far to do so?

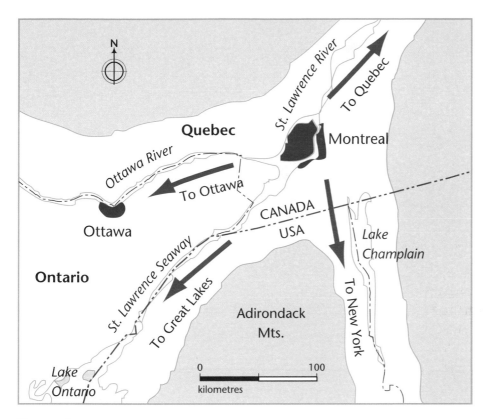

Figure 3
Montreal is a route centre.

Satellite Images

Satellite images are used to study not only physical geography and the environment, but also cities. Satellite images are not photographs. Rather, they are images created from the digital signals transmitted from satellites. These signals show the amount of radiation coming from the Earth. From the satellite data, images like the one shown in Figure 4 on page 42 can be created. This image uses false colour to show more detail. The urban core is purple and pink. Residential areas show as pale blue.

Figure 4
A Satellite Image of Winnipeg

Case Study   *The Growth of Winnipeg*

Often, a city grows because a number of factors interact. The city's situation as a route centre affects historical events. These events make the city's situation even better. As a result, the city grows even more. Winnipeg is an example of such interaction.

In Figure 4, "X" marks the place where the Red River is joined by its tributary, the Assiniboine River. At this site, Fort Garry, a Hudson's Bay Company trading post, was established in 1822. In 1873, the settlement that had grown around the trading post was incorporated as the city of Winnipeg.

Ten years later, the Canadian Pacific Railway was pushing westward across Canada and reached the Red River. The railway had to cross the river at some point between the United States border and where the river empties into Lake Winnipeg 165 km further north. The obvious crossing point was at the small city of Winnipeg. In Figure 4, "Y" marks the spot where the railway bridge was built.

The railway instantly made the city more important. At the same time, the production of Manitoba wheat for export was expanding. To transport the wheat, a network of railways was built, all of them focusing on Winnipeg. In Figure 4, the two "Z"s mark large railway marshalling yards. These historical events led to the rapid growth of the city. Its population increased to provide goods and services for Manitoba and beyond. Today, the people of Winnipeg can thank both the city's situation and historical events such as railway construction for the city's growth.

Discover For Yourself

1. In your own words, explain the difference between Winnipeg's *site* and *situation*.
2. Divide into small groups to study Figure 4. Identify
   a) the urban core of the city
   b) Winnipeg International Airport
   c) the Greater Winnipeg Floodway that surrounds the city
   d) evidence that shows that the land is flat

Turn to page 259 to learn more about analyzing satellite images.

WEB LINK To view other satellite images of Canada, look up
http://www.ccrs.nrcan.gc.ca/ccrs/imgserv/tour/toure.html
Click on map locations to see the images.

Urban Growth Around the World

In the year 2000, almost half of the world's six billion people were urban. Geographers predict that 60 per cent of the world's population will be urban by the year 2015. Figure 5 shows the steady rise in urban population over the past half century.

Geographers use the term **urbanization** to describe the rapid growth of cities. They study the reasons for urbanization in different regions of the world. They have discovered that today's patterns of urbanization in *developed* countries, such as Canada and the United States, are different from the patterns in *developing* countries, such as those in Africa. Before we look at these patterns, we need a better understanding of the differences between developed and developing countries.

urbanization—the growth of cities. This term can also mean the adoption of an urban lifestyle.

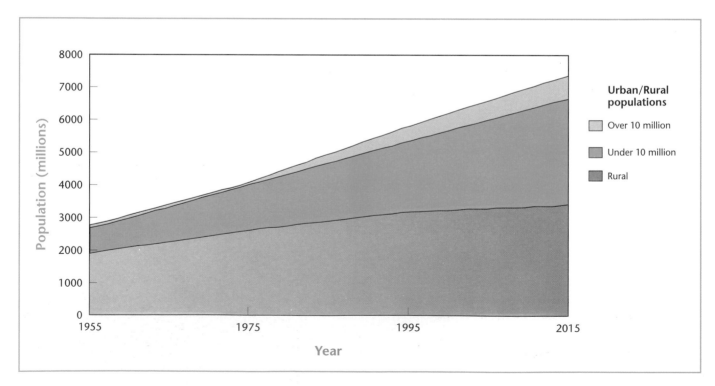

Figure 5

The Rise in World Rural and Urban Population Since 1955

Developed and Developing Regions

To gain a better understanding of the human world, geographers look for ways of dividing the world's countries into groups based on specific common features. One of these groupings creates *developed* and *developing* regions. Developed countries are industrialized. The average income of their populations is generally high. They range from the richest country in the world (the United States) to less wealthy countries, such as Portugal. Developing countries depend more on agriculture (although some, such as Korea, are industrializing rapidly). In some developing countries (for example, in certain African nations), people's incomes and living conditions are very poor. The chances that these conditions will improve are also poor.

Patterns of Urbanization

In much of the developing world, many people are poor and live from hand to mouth. That means that they struggle to meet their basic needs of food, clothing, and shelter, as well as to gain access to health care, education, and employment. Rural areas especially can be desperately poor. The prospect of getting even a part-time job, very basic health care, or more than the lowest levels of education encourages millions of people to move from the countryside to cities. This has led to rapid urban growth in developing countries. People flock to cities and often stay in makeshift shantytowns, making a living any way they can.

In the developed world, rural-urban migration began on a large scale during the Industrial Revolution. By the late 20th century, this trend had tapered off, so that some cities in developed countries are now actually losing population. Some people in North America now seem to prefer a rural lifestyle. Cars give them the freedom to drive into the city as often as they want.

Figures 7 and 8 show the patterns of urbanization in developing and developed regions.

Figure 6
Rural-urban migration has resulted in great increases in the populations of cities in the developing world. City housing is limited, so people must build *shantytowns*.

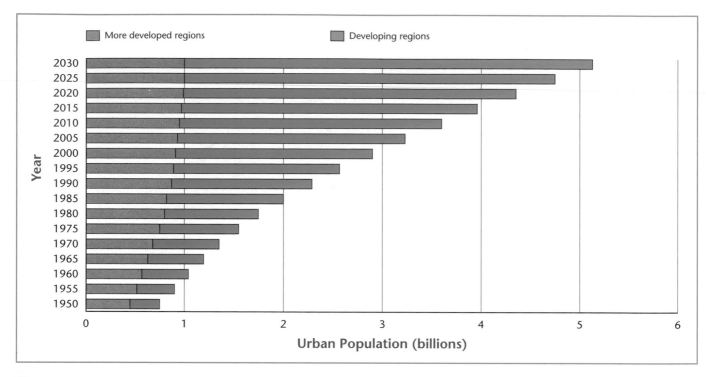

Figure 7
Urban Population in Developed
and Developing Regions

 **With Maps**

1. According to Figure 8,
    a) which continent has the lowest percentage of urban
       population?
    b) which parts of the world have the highest percentages of
       urban population?
2. Figure 9 is an organizer that can help you compare countries on
   the basis of the percentage of their populations that live in cities.
   Divide into small groups to fill in this organizer. Make a group
   copy of the organizer and then follow these steps.
    a) In column (a), list four countries that have less than 25 per
       cent of their population living in cities. Use Figure 8 and an
       atlas to help you.
    b) Discuss what you believe is the level of development in the
       country. When you have reached a conclusion, write in
       "developed" or "developing" beside each country.

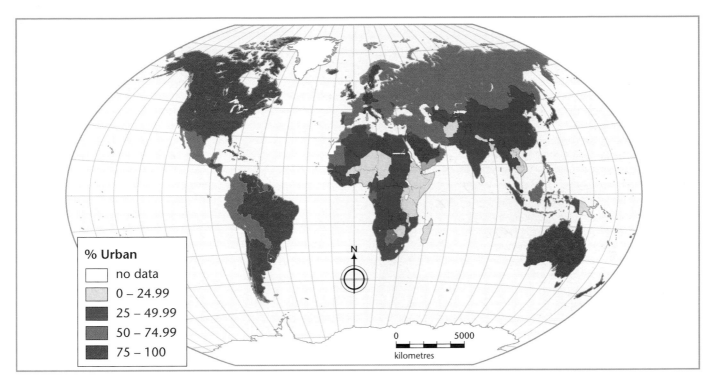

Figure 8
The Percentage of the Population that is Urban Dwelling in Countries Around the World

c) Follow the same steps to fill in the rest of the columns in the organizer.

(a) Under 25% urban	Developed or developing?	(b) 25% – 75% urban	Developed or developing?	(c) Over 75% urban	Developed or developing?

3. Describe and explain any connection you think there may be between the level of development of a country and its percentage of urban population.

Figure 9
You may need an atlas to help you fill in this chart.

The World's Largest Cities

The populations of some cities in the world have grown to over 10 million people. These "mega-cities" have developed because they perform many different functions. For example, a large city may be

- ▸ A major port
- ▸ A political capital
- ▸ A world financial centre
- ▸ A historic city with royal connections
- ▸ A railway centre
- ▸ A tourist and entertainment centre

Some cities, such as London in the United Kingdom, are *all six* of the above. It is not surprising, then, that London was the first city in the world to reach a population of 10 million.

In the 20th century, several cities reached and overtook London in size. After the Second World War, Japan's rapid economic recovery helped make Tokyo both the world's most populated city and the largest manufacturing city in the world. Most homes in Canada and the United States have electronic products manufactured in Tokyo and other leading Japanese cities.

The developing world has many rapidly growing mega-cities. All of them are experiencing a tidal wave of rural-urban migration. Mexico City has grown to become the largest city in Latin America, closely followed by São Paulo in Brazil.

Many problems result from the rapid growth of cities in poorer countries. Overcrowding leads to large slum areas that lack adequate housing and sanitation. There are not enough full-time jobs. In some cities, children roam the streets trying to earn small amounts of money by shoe-shining or looking after parked cars. Crime is often a serious problem. Also, environmental controls are very limited when industrial growth occurs. As a result, air pollution can be extreme in places such as Beijing and Mexico City.

With rapid urban growth in Asia, Latin America and even Africa (for example, Lagos, Nigeria), the cities of North America and Europe are falling behind in the world ranking of mega-cities.

Figure 10
London, a port city situated on the Thames River in England, was the first city in the world to reach a population of 10 million.

Figure 11
Air pollution, visible in the smog hanging over Mexico City, is just one of the problems resulting from rapid urban growth in developing countries.

 **For Yourself**

1. Figure 12 on page 50 shows the past and expected future growth of the world's largest cities. Get or make an outline map of the world and mark and label the 16 large cities listed. Use two different sizes of symbols (for example, squares or triangles) to mark them: one size for cities with 1996 populations of *over* 15 million, and another size for cities with 1996 populations of *under* 15 million.

2. Which city is projected to grow the *least* between 1970 and 2015? What reasons can you think of for this?

3. Which city is projected to grow the *most* between 1970 and 2015? What does this tell you about urbanization in the continent the city is located in?

MATH
LINK

WEB LINK

To find out more about the populations of the world's cities, look up http://www.un.org/Depts/unsd/demog/index.html Click on the region of the world map you are interested in.

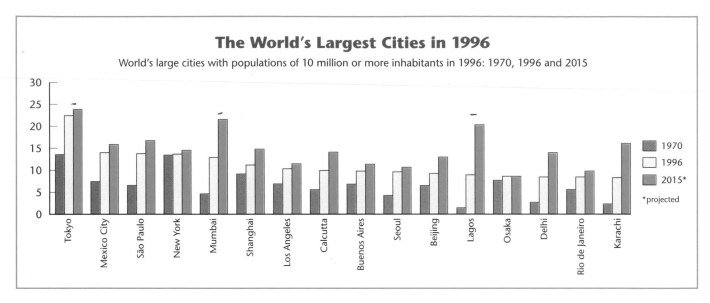

The World's Largest Cities in 1996

World's large cities with populations of 10 million or more inhabitants in 1996: 1970, 1996 and 2015

Legend:
- 1970
- 1996
- 2015*

*projected

Figure 12
Mega-cities of the World, 1996

Summary

In this chapter you have discovered how site and situation have affected the growth of cities. You have studied the factors that have led to urbanization in different parts of the world. You have also learned that cities in the developing world have been growing more quickly than cities in the developed world.

Reviewing Your Discoveries

1. a) Describe one advantage and one disadvantage of living in the countryside.
   b) Describe one advantage and one disadvantage of living in a city.
2. List and describe three differences between cities in the developed world and cities in the developing world.
3. Imagine that you are a member of a poor rural family in an African country. Write a letter to a relative or a diary entry to yourself explaining why your family is planning to move to a large city.

Using Your Discoveries

1. Make a copy of the organizer in Figure 13. Fill it in with a summary of the features of Canada's top six cities. Use an atlas, the information in this chapter, and the Internet to help you.

City	Population	Site	Situation	Function
Toronto				
Montreal				
Vancouver				
Ottawa				
Edmonton				
Calgary				

Figure 13
Organizer for Question 1

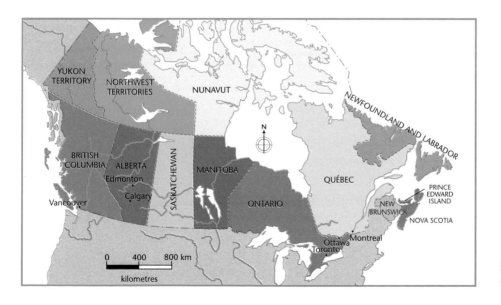

Figure 14
Canada's Six Largest Cities

2. Research the geographical factors that led to the siting and growth of your community. Use both library resources and local experts to help you.

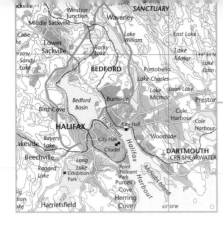

Chapter 4

Patterns in Cities

Key terms

landscape

central business
district (CBD)

utilities

landscape—what we see when
we look around.

In this chapter we focus on land use patterns in cities.
The information and activities will help you

▸ identify and describe different types of land use
▸ identify some issues resulting from changes in workplaces
▸ show your understanding of some Canadian employment
patterns.

A Centre for Activities

Think about the places you use in your nearest town or city. A list
of them would include the places in which you live, shop, attend
school, borrow books, get medical treatment, play sports, see
movies, and so on. Your list would also include the places along
which you travel to and from your activities. This vast collection
of places, arranged in patterns on the city's land, creates different
types of urban **landscapes**.

Geographers look for patterns in the way urban landscapes are
arranged. For example, how are *residential* landscapes (houses and
apartment buildings) arranged in relation to *industrial* landscapes
(factories and warehouses)? They divide urban landscapes into six
main types:

▸ residential (houses and apartment blocks)
▸ commercial (offices and shops)
▸ industrial (factories and warehouses)
▸ institutional (hospitals, schools, and other public buildings)
▸ transportation (canals, railways, roads, and airports)
▸ recreational (sports fields, recreation centres, golf courses,
and so on)

Figure 1
Label and describe each of these urban landscapes. What types of jobs do people do in each of them?

Land Use Maps

Land use maps show how land in an area is divided up for different uses. City land-use maps identify the areas shown in the legend below. (Large-scale land use maps also show institutional buildings such as schools and hospitals.)

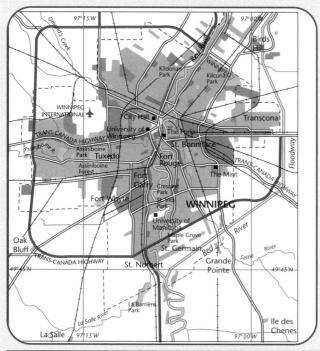

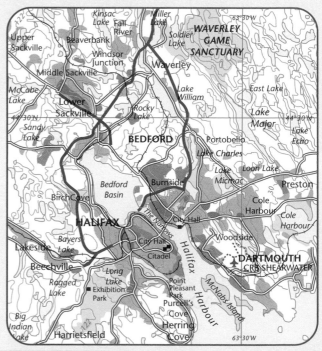

LEGEND

Boundaries

- - - - - county/municipal/district/city

Physical features

~~~ river

 contours

• 155 spot height in metres

Land use

◻ central business district

◼ other major commercial areas

◼ industrial

◼ residential

◼ major parks and open spaces

◻ non-urban

Communications

━━━ expressway/multilane highway

═══ other highway

┼┼┼ major railway

┄┼┄ canal

✈ airport

Figure 2
Land Use in Winnipeg

Figure 3
Land Use in Halifax

 With Maps

1. Look at the Winnipeg land use map in Figure 2.
   a) Which types of land use occur nearest to rivers and railways?
   b) Compare the Winnipeg land use map with the two models of urban land use in Figure 4. Which model does Winnipeg follow? Explain your answer.
2. In Figure 3, how does the physical geography of Halifax affect the layout of residential and industrial areas?

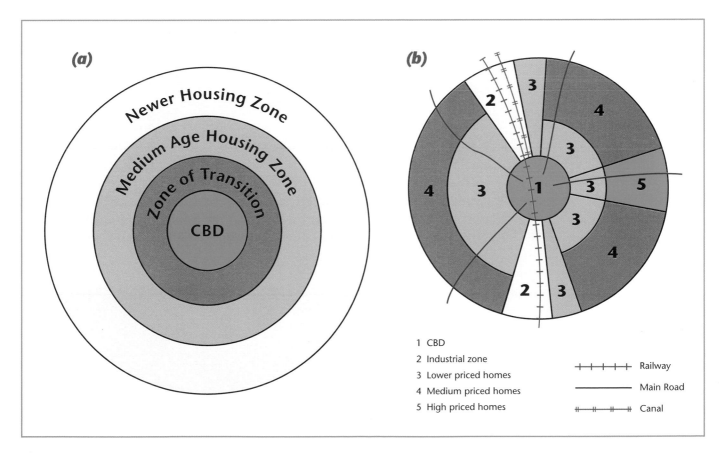

Figure 4
Some geographers say that cities are arranged in a series of *circles* from the city centre outward (a). Others say that cities are arranged in *wedges* following transport routes from the city centre outward (b).

The Central Business District

A separate type of land use shown in Figures 2 and 3 is the **central business district** or **CBD**. In your town or city, you might refer to the CBD as "downtown" or "the city core." You might spend time in the CBD at its department stores and movie theatres, and you may know adults who work in the CBD's office buildings.

Land in the CBD is expensive and many people compete for it. Buildings are usually tall, as in Figure 1(a) on page 53, to fit as many businesses and people as possible into an area. Organizations such as banks, department stores and theatres can most easily afford to pay the steep prices of CBD land. Other groups, such as house builders and factory owners, usually seek cheaper locations with more space.

Residential Land Use

Since the CBD is a place of business and entertainment, few people actually live there. Most residential areas are outside the city centre. They often form three zones of housing with the characteristics described in Figure 6. The different densities in the three zones are related to the price of land. Land is more expensive in Zone 1, so the best way to house those who want to live close to the city centre is to build high-density apartment buildings.

Figure 5
Land is cheaper in the suburbs, so homes and lots are larger.

Zone	Location	Housing	Population Density
1	– just outside CBD	– older housing – many apartment buildings	high
2	– further from CBD	– medium age (built 20 to 60 years ago)	medium
3	– outer suburbs	– recent (built since 1970s) – bigger homes on larger lots	low

Figure 6
Housing Zones and Population Density in Relation to Distance from the Central Business District

 **With Maps**

1. Figure 7 on page 58 is a census tract map of Ottawa.
   a) Describe the population density pattern near the centre of Ottawa.
   b) Describe the population density pattern in the outer suburbs.
2. Figure 8 on page 59 is a satellite image of downtown Vancouver and Vancouver's harbour. "X" marks English Bay, a zone of transition made up of high-rise apartments. English Bay has the highest population density of any urban area in Canada (24 000 people per km^2). "Y" marks the city centre.
   a) Give two reasons why English Bay is a popular place to live.
   b) Using an atlas or reference book to help you, match each of the letters "A" to "F" on Figure 8 to the features listed below.
      – Stanley Park
      – False Creek
      – Lions Gate Bridge
      – Railway yards
      – BC Place
      – Yacht or sailboat marinas

Turn to page 21 to review Census Tracts.

Figure 7
Population Density by Census Tract Map of Ottawa

Figure 8
Satellite Image of Part of Vancouver

Commercial Land Use

People work, buy goods and use services in a city's commercial areas. The most common types of jobs in commercial areas are in offices and stores. In Canada, the wholesale and retail trade is the largest single type of service employment in cities.

In the past, people wanting to shop in the city were limited to the CBD's stores. This changed as cities grew and residential areas spread outward. It would be too time-consuming for suburban residents to drive frequently to the CBD. They would tire of the traffic congestion and expensive parking.

For these reasons, *regional shopping centres* were built in the suburbs. These are large shopping malls, like the one shown in Figure 1(c). They have lots of parking space for suburban shoppers, who use their cars often. These malls usually feature one or more department stores, as well as many clothing stores, specialty stores (for example, jewellers), services (for example, banks and hairdressers) and a food court.

District shopping centres serve a smaller area than regional shopping centres. They usually do not have a major department store, but may feature a large hardware or general purpose store, as well as a more limited variety of stores and services. *Neighbourhood shopping centres* contain only a few shops and serve the local area. Below this level, there are many corner stores providing mainly convenience foods.

An interesting pattern has developed on commercial land: the same types of shops and services often locate near each other. For example, fast food shops in malls are grouped together in food courts, rather than being scattered throughout the mall. Fast food shops are also near each other in outside strip malls. Often, several electronics stores are found in the same area of a city. There are many advantages to this pattern. For the consumer, the advantage is easy comparative shopping, since the same kinds of stores are right next to each other. For storeowners, the advantage is a higher number of customers, since people prefer to shop in an area where they can compare prices. This advantage to storeowners seems to outweigh the main disadvantage: having to compete with neighbouring stores for customers.

Figure 9
The owners of these three record stores may have been hoping to attract comparison shoppers and a larger number of customers by locating next to each other.

 **For Yourself**

1. Figure 10 shows how people are employed in four Canadian census metropolitan areas (CMAs). The numbers represent the percentage of the labour force in each industry group. Divide into groups of four to analyse the information in this figure. Follow these steps.

Industry group	Toronto (%)	Ottawa (%)	Hamilton (%)	Saskatoon (%)
Primary (farming, forestry, mining, fishing)	0.7	1.2	1.8	5.9
Manufacturing	16.8	6.6	20.3	9.5
Construction	4.9	4.8	5.3	5.4
Transport, Communications & Utilities	7.3	6.3	5.9	7.8
Wholesale & Retail Trade	18.2	14.2	18.4	17.9
Finance, Insurance & Real Estate	8.9	4.9	6.0	4.7
Government Services	3.8	19.7	4.0	5.5
Educational, Health & Social Services	14.5	17.5	18.4	21.2
Business Services	10.7	10.3	6.4	5.3
Other Services	14.2	14.5	13.5	16.8
Total	100.0	100.0	100.0	100.0

Figure 10
Employment in Four Canadian Census Metropolitan Areas (CMAs)

a) List the industry groups that could be employed in a city's commercial areas (the CBD or at regional, district or neighbourhood shopping centres). Give examples to support the choices on your list.

b) List, in order, the three industry groups that employ the most people. This can be figured out by adding each group's percentages across the four CMAs.

c) From what you already know or can find out, why do you think Toronto has an unusually high percentage of workers in finance, insurance and real estate?

d) From what you already know or can find out, why do you think Ottawa has an unusually high percentage of workers in government services?

e) From what you already know or can find out, why do you think Hamilton has an unusually high percentage of workers in manufacturing?

f) From what you already know or can find out, why do you think Saskatoon has an unusually high percentage of workers in farming/forestry/fishing/mining?

g) Make four circle graphs (one graph per group member) to show the employment patterns of the four cities in Figure 10. Each circle graph will have 10 sections. Multiply each industry group's percentage figure by *3.6* to give you the number of degrees (out of 360) that each section should take up in the circle's circumference.

2. Describe the shopping centre nearest to your home or school.
a) Is it in the CBD or in the suburbs?
b) If it is in the suburbs, is it a regional, district, or neighbourhood centre? Explain your answer.

Turn to page 260 to review making circle graphs.

Industrial Land Use

There are two typical locations for industrial activities in cities. One is *near rail or water transport*. This location is necessary for industries that use raw materials in bulk, such as steelmaking, oil refining and grain milling. Grain milling and oil refining are often located on *estuaries* (wide river mouths) where their raw materials are easily imported.

The second location is for industries that depend on a good road network and need space for their factories. These industries are usually located *in the suburbs of cities, close to freeways or airports.*

Figure 11
Where a company locates depends partly on the product it produces. Why do you think the plant in (a) chose to locate on a river, while the factory in (b) situated itself next to a highway?

Discover For Yourself

1. Look at the two industrial scenes shown in Figure 11. Which of the industries shown do you think deals with the import of raw materials? Which one deals with the distribution of a finished product? Explain your choices.

2. In small groups, make a list of factories and workshops in your town or city. Divide up the factories and workshops among the group members. Each member should research the answers to these questions about his or her factories. Summarize your answers in an organizer.

   a) Where is each factory located? Is it near rail or water transport, or in the suburbs?

   b) Why is the location of each factory a good one? Include information about the factory's raw materials, main form of transport, or main products.

Recreational Land Use

Parks and other open spaces give people in the city a chance to relax and enjoy nature. Most cities also set aside land for people who take part in *active recreation*. Baseball diamonds, tennis courts, golf courses, and other sports fields are provided on this land. For winter recreation, Canadian cities provide indoor sports facilities for basketball, ice sports, and club activities for all ages.

1. Make a list of four indoor recreational activities (for example, going to the movies) and four outdoor recreational activities (for example, playing soccer) that you regularly take part in. Fill in information about each one under the headings in Figure 12.

List of Activities	Number of visits per week	Location (town centre, local or elsewhere)	Means of transport	Type of building or ground
Indoor				
1				
2				
3				
4				
Outdoor				
1				
2				
3				
4				

Figure 12
Recreation Activity Organizer

2. In small groups, discuss the quality of recreational facilities in your local residential area. Are there enough? Do they fit in with the landscape? Do they offer enough activities at convenient times? Are they large enough to meet the needs of their users? Come up with a list of suggestions for how to improve your area's recreational facilities.

Using Land for Movement

A great deal of movement is constantly taking place in cities. Some movement, such as much of the **commuting** done to school and work, uses land above ground on busy and noisy transport routes. About one quarter of all space in cities is taken up by roads, parking areas, railways, and other means of transport.

The movement of services and information often takes place beneath the land surface, through quiet, high technology systems. Telephone service is one example. For many years, the telephone system made use of overhead wires attached to poles on the land. Today, telephone wires come to many homes underground, which helps to improve the appearance of the residential landscape. Figure 13 shows the journey chain of one of the other **utilities** that are delivered underground to users.

commuting—the daily movement to or from a place of work or study.

utilities—things that are useful to us. City utilities include gas, water, electricity, cable, and telephone systems.

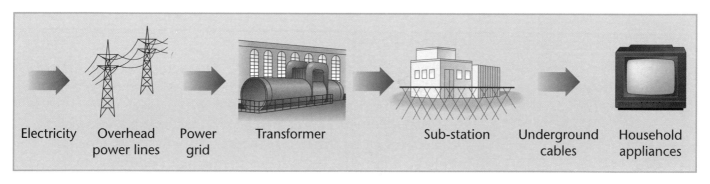

| Electricity | Overhead power lines | Power grid | Transformer | Sub-station | Underground cables | Household appliances |

Figure 13
Electricity is supplied to overhead power lines from hydro, coal-fired or oil-burning power stations. This figure shows its journey from overhead lines to home users.

WEB LINK

To see fascinating diagrams of the utilities and services located underground in New York City, look up http://www.national geographic.com/nyunderground/docs/nymain.html

Changes in City Workplaces

Along with telephone wires, fibre optic cables are now being connected underground. They provide cable TV service as well as high-speed access to the Internet. The Internet, electronic mail (e-mail), faxes and modems are some of the technologies that are changing city workplaces. Another one is *teleconferencing*, which enables people to talk on the telephone with a number of others at the same time.

These and other changes mean that many people no longer need to commute to an office. Employees can use their computers to send and receive work as well as communicate with their colleagues electronically. This is called *telecommuting*, and some sociologists predict it will become very common in the future in developed nations such as Canada. As more employees work at home, urban landscapes are bound to be affected. Residential, commercial, and transportation land use will have to adapt to these changes.

Urban Planning

Does your city have any problems? Are there any older, unattractive places that people describe as "eyesores"? Are there traffic or road problems? Do more jobs need to be created?

Urban planners are the people who try to solve some of the problems of cities. These problems have two causes. One is unplanned growth. The other is the aging of buildings and roads in the older city centre. The recommendations planners make to solve these problems are voted on by elected councils.

One way to improve city life is to divide the city's land into *zones* of particular uses (residential, industrial or commercial). Residential zones can then be placed next to open space and far away from the pollution of industrial zones.

Another goal of planners is to reduce traffic congestion. Solutions include one-way street systems and pedestrian-only areas. The overall volume of traffic and pollution can be reduced by a well-planned and widely used public transit system.

In an attempt to keep city centres vibrant and active, many major cities around the world have undergone redevelopment to make themselves more people-friendly. Sometimes planners focus on improving a very small area in a city. One such project is the Yonge-Dundas Redevelopment Project in Toronto. This project involves the creation of an entertainment centre. A one-acre site on one corner will include a major subway entrance, canopied spaces, fountains, and a stage for public performances. Across the street will be a large movie theatre, a hotel, and shops. Some of the centre's cinemas will be used during the day by the nearby Ryerson Polytechnic University. It is hoped that the centre will attract people as a meeting place. These people may then become customers of local businesses.

Figure 14
Plans for the Yonge-Dundas Redevelopment Project in Toronto

Discover  *With Maps*

1. Figure 15 is a plan of Edmonton. It shows, in a general way, how land in Edmonton might be developed in the future.
   a) Why do you think the transportation corridor has been planned in the shape of a ring around the city of Edmonton?
   b) What is the advantage of having industrial areas in large blocks to the northwest and southeast of the city?
   c) Why do you think there is no development planned along the north Saskatchewan River?
2. Imagine you are a planner for your town or city.
   a) Make a list of the problems you would try to solve.
   b) Make a list of the strategies you would use to solve these problems.
   c) Would any of your strategies create further problems?
   d) Compare your lists with your classmates'. What are the similarities and differences between them?

Summary

In this chapter you have learned about the many activities people do in towns and cities. You have seen how these activities create landscapes, arranged in various land use patterns. You have also discovered that urban planners are needed to work on the common problems of cities.

Reviewing Your Discoveries

1. Sketch a likely scene for each different kind of urban landscape (residential, commercial, industrial, institutional, transportation, recreational).
2. List the reasons why banks and offices concentrate in the central areas of cities.
3. Make a web to show all the ways in which computer technology may change cities.

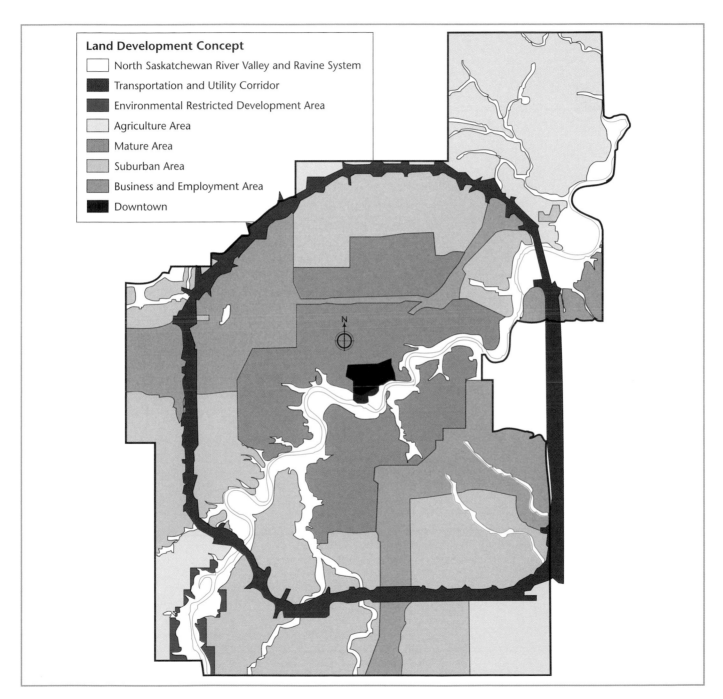

Land Development Concept

☐ North Saskatchewan River Valley and Ravine System
■ Transportation and Utility Corridor
■ Environmental Restricted Development Area
☐ Agriculture Area
☐ Mature Area
☐ Suburban Area
☐ Business and Employment Area
■ Downtown

N

Figure 15
A Municipal Development Plan for Edmonton

Using Your Discoveries

1. Imagine that a city's elected council bans cars from entering the city's central area. In small groups, discuss these questions:

   a) What would be the advantages of this ban?

   b) What would be the disadvantages?

   c) What is your personal stand for or against the ban? Debate your position with someone with an opposing point of view.

2. Arrange a field trip to the CBD of a city with a population of over 10 000 people. It could be in your own city or in one near by. Do a *pedestrian count* to try to find the centre, or busiest spot, in the CBD, where land would be the most valuable. Follow these steps.

   a) Divide up into pairs. Each pair should make a tally sheet like the one in Figure 16.

   b) Working in pairs, spread out around the CBD. One member of each pair will count all the people that pass by, including children. The other member will record each passerby on the tally sheet.

   c) Decide how long you will count people (e.g., for a five or ten minute period) and record your data.

   d) Compare your tally sheets. The highest pedestrian density is likely to be at or close to the centre of the CBD.

Location _____ Date _____ Time _____

	0–5 mins.	5–10 mins.	10–15 mins.	15–20 mins.	20–25 mins.	25–30 mins.	Total
Left to Right	~~HHH~~ III						
Right to Left	~~HHH~~ ~~HHH~~ II						

Figure 16
Tally Sheet for the Pedestrian Count

Chapter 5

Population Growth

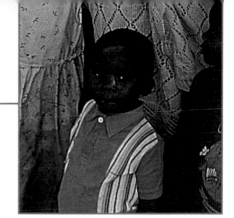

In this chapter we focus on patterns in population growth.
The information and activities will help you

▸ describe population characteristics using the correct terms
▸ show your understanding of relationships between
  population characteristics
▸ analyze and make population pyramids.

Key terms

O━┱

population pyramid
birth rate
death rate

One Minute in the Life of the World

Time a minute using a watch or clock. As you count the seconds,
make a guess at the number of babies being born.

When the minute is up, the world's population will have
about 260 new babies. During the same minute, about 100 people
will have died. That makes a
gain of about 160 people *every
minute*. The gain *every hour* is
about 9600 people. What do
you calculate the increase to
be in *one day*? What about in
one year?

All of these added children
have to be fed, clothed and
housed. When they grow up,
they will need jobs to earn a
living. Providing these needs for
so many added people may be
the greatest challenge of the
new millennium.

Figure 1
How many seconds does it take
for the number of babies shown
here to be added to the world's
population?

Age Patterns

Is it possible to predict patterns in a country's population growth? What information would you need to do so?

One kind of information is the age and sex of each citizen. This is provided in a country's census. Countries with many young people will have higher growth rates in the future than countries with many seniors. Countries with many women able and willing to have children will have higher growth rates than countries with fewer women prepared to have children. The best way to show the age/sex pattern of a country is to construct a **population pyramid**.

population pyramid—a set of two bar graphs placed back to back against a vertical axis. One shows the numbers of males, and the second shows the numbers of females, in different age groupings in a country.

Discover With Graphs

1. Look at Figure 2.
   a) Find which bar in the pyramid you would fit into if you lived in Mozambique.
   b) What are the three large age groupings shown by the three different colours? (These are groupings used by the United Nations.)
   c) Match each large age grouping to one of these descriptions:
      – people mainly too young to work
      – people of retiring age
      – people of working age
2. Another larger grouping that interests geographers is "females between 15 and 45." This grouping describes women of child-bearing age. Why is it helpful to know the number of people in this age/sex group?
3. Use the data in Figure 3 on page 74 to draw a population pyramid for Canada in 1996. To do so, use only the *percentage* figures for every age group. Make your scales match the scales on Figure 2 on page 73 (1 cm for every 1 per cent on the horizontal scale, and 0.5 cm for each age group in the vertical scale). Complete your graph with an appropriate title.

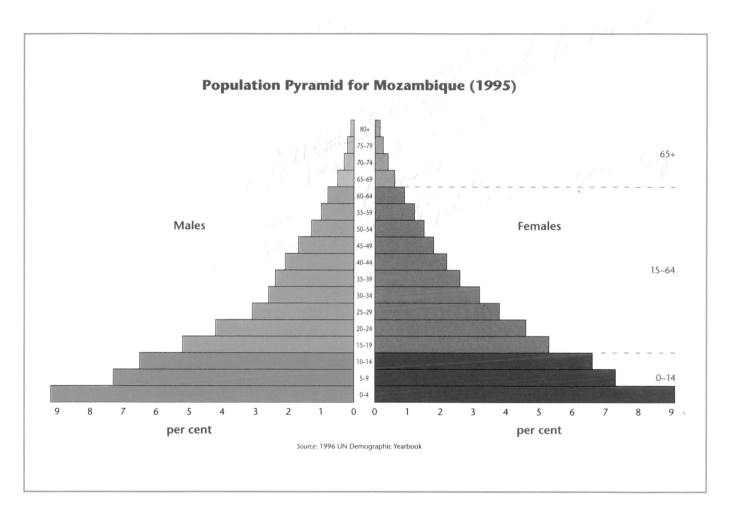

Population Pyramid for Mozambique (1995)

Males

Females

65+

15–64

0–14

per cent

per cent

Source: 1996 UN Demographic Yearbook

Figure 2
This is a population pyramid
for Mozambique in 1995.
Mozambique is a developing
nation in south-east Africa.

Turn to page 262 for more information on making population pyramids.

4. Note the difference in shape of the pyramids for Canada and Mozambique. For both Canada and Mozambique, add up the percentages shown by the bars to find the total percentage of people who are
   a) under 15 years of age
   b) of working age (15–64 years)
   c) over 65 years of age

Ages	Males (in 000s)	Male (%)	Females (in 000s)	Female (%)
85+	109.5	0.4	249.9	0.8
80–84	174.8	0.6	292.8	1.0
75–79	289.3	1.0	415.6	1.4
70–74	433.9	1.4	547.5	1.8
65–69	536.9	1.8	593.4	2.0
60–64	597.1	2.0	617.5	2.1
55–59	662.4	2.2	670.8	2.2
50–54	838.5	2.8	834.1	2.8
45–49	1085.2	3.6	1074.7	3.6
40–44	1192.5	4.0	1196.4	4.0
35–39	1344.9	4.5	1323.3	4.4
30–34	1335.0	4.5	1298.3	4.3
25–29	1122.3	3.7	1103.1	3.7
20–24	1033.9	3.5	1003.5	3.3
15–19	1026.7	3.4	977.0	3.3
10–14	1032.3	3.4	988.1	3.3
5–9	1031.8	3.4	984.9	3.3
0–4	1000.2	3.3	951.1	3.2
Total	14847.2	49.5	15122.0	50.5

Figure 3
Canada's Population in 1996 by Sex and Five-Year Age Groups

5. In small groups, use your answers to Question 4 to discuss these questions:
   a) What challenges does the Mozambique government face in providing education?
   b) What are the prospects for future population growth in Mozambique?
   c) What challenges does the Canadian government face in providing pensions for seniors?

6. Your pyramid shows that Canada in 1996 had a large number of people aged between 30 and 50. This group of people is called the **baby boom** generation, or "baby boomers." This term means that an unusually high number of births occurred 30 to 50 years earlier.
   a) Compare the percentage of people in Canada in 1996 aged between 10 and 29 with the percentage of baby boomers. Is it greater or less?
   b) Based on your comparison, when would you say the baby boom ended (that is, around which year did the number of births start to fall, so that there were fewer people in certain age groups in 1996)?

baby boom—a period of about 20 years, following World War Two, during which there were an unusually high number of births. It was followed by the "baby bust"—a period of significantly less births.

Figure 4
The baby boom led to a housing boom in the suburbs in the 1950s.

WEB LINK
To see examples of population pyramids (including animations) for any country, look up
http://www.census.gov/ipc/www/idbpyr.html

Population Change

Two important factors in population change are *births* and *deaths*. A country's birth and death rates measure the number of births and deaths per thousand people. They are used to compare population patterns in different countries. For example, the United Kingdom has a higher death rate than Brazil. In this case, the different death rates tell us that the United Kingdom has an older population than Brazil. Figure 5 shows the world's birth rate patterns. Figure 6 shows the world's death rate patterns.

Figure 5
Birth Rates Around the World, 1997

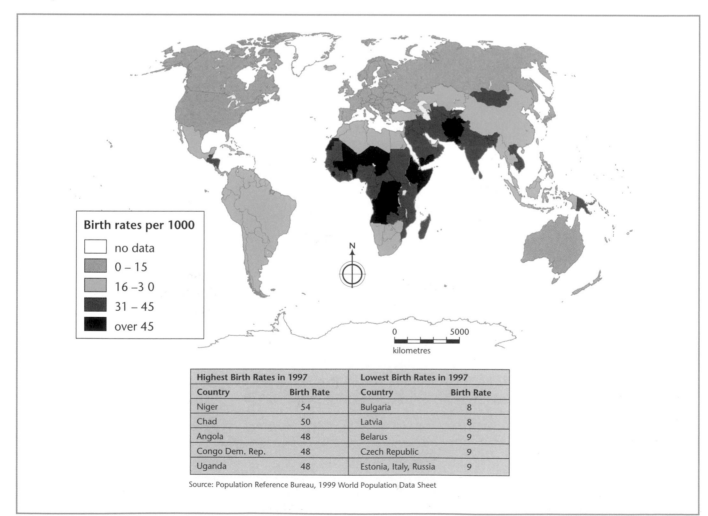

Birth rates per 1000

	no data
	0 – 15
	16 – 30
	31 – 45
	over 45

Highest Birth Rates in 1997		Lowest Birth Rates in 1997	
Country	Birth Rate	Country	Birth Rate
Niger	54	Bulgaria	8
Chad	50	Latvia	8
Angola	48	Belarus	9
Congo Dem. Rep.	48	Czech Republic	9
Uganda	48	Estonia, Italy, Russia	9

Source: Population Reference Bureau, 1999 World Population Data Sheet

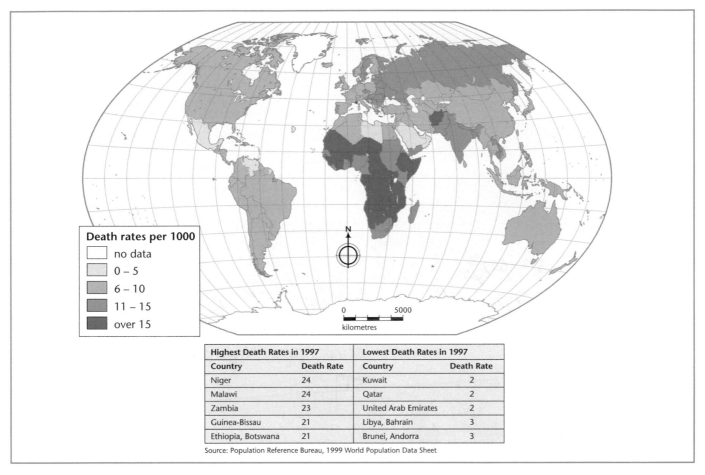

Highest Death Rates in 1997		Lowest Death Rates in 1997	
Country	Death Rate	Country	Death Rate
Niger	24	Kuwait	2
Malawi	24	Qatar	2
Zambia	23	United Arab Emirates	2
Guinea-Bissau	21	Libya, Bahrain	3
Ethiopia, Botswana	21	Brunei, Andorra	3

Source: Population Reference Bureau, 1999 World Population Data Sheet

Figure 6
Death Rates Around the World, 1997

When geographers study population change in a country, they are not only concerned with births and deaths. They also need to know how many *immigrants* come into a country and add to its population. Likewise, they count the number of *emigrants* who leave the country and lower its population.

Let's summarize these factors. If we know the population of Canada in May of 2001 (the month of the census), the population in May of the year 2002 will be

The population in May 2001
+ Births between May 2001 and May 2002
– Deaths between May 2001 and May 2002
+ Immigrants to Canada between May 2001 and May 2002
– Emigrants from Canada between May 2001 and May 2002.

Discover *With Graphs and Tables*

birth rate—the total number of births per thousand people in a country's population. The formula for calculating it is **Total Births ÷ Total Population × 1000**.

death rate—the total number of deaths per thousand people in a country's population. The formula for calculating it is **Total Deaths ÷ Total Population × 1000**.

1. A birth rate over 30 is considered high. Mozambique's birth rate is 41.
   a) Look back at Mozambique's population pyramid (Figure 2 on page 73). Describe its shape.
   b) Why would countries with this kind of population pyramid tend to have high birth rates?

2. Divide into groups of four. Make a group copy of the organizer in Figure 7. Each member of the group should choose two different countries. For your two countries, fill in the **birth rate** column of the organizer by following these steps:
   a) Divide the total number of births in a country by the country's total population.
   b) Multiply the number in a) by 1000.

3. For your two countries, fill in the **death rate** column by following these steps:
   a) Divide the total number of deaths in a country by the country's total population.
   b) Multiply the number in a) by 1000.

4. For your two countries, fill in the last two columns by following these steps:
   a) Calculate the natural increase by subtracting the number of deaths from the number of births.
   b) Calculate the *rate* of natural increase by subtracting the death rate from the birth rate. Then divide by 10.

5. As a group, fill in the missing columns for the world in the last row. Then study your results. Which two countries are making the largest contribution to the world's total population?

6. Calculate the population of Canada in 2002 by assuming:
   a) The population in 2001 is 30 700 000.
   b) During the 2001–2002 period, there were
      – 355 000 births
      – 215 000 deaths
      – 250 000 immigrants
      – 30 000 emigrants

Country	Population mid-1999 (in 000s)	Births (in 000s)	Birth Rate (per 000)	Deaths (in 000s)	Death Rate (per 000)	Natural Increase (in 000s)	Rate of Natural Increase (%)
Canada	30 600	337		214			
United Kingdom	59 400	715		595			
Japan	126 700	1 265		890			
China (inc. H.K.)	1 261 000	20 200		8 800			
India	986 600	27 625		8 885			
Kenya	28 800	1 010		403			
Brazil	168 000	3 530		1 010			
Mali	11 000	517		176			
World	5 982 000	137 590		53 850			

Figure 7
Births and Deaths for Selected Countries in 1999

Lowering Birth Rates

There is another way of describing a country's birth patterns. The *total fertility rate* measures the average number of births a woman has in her lifetime. A total fertility rate of about 2.1 children per woman is necessary for a population to replace itself over the long term. In Niger (birth rate: 54) and Ethiopia (birth rate: 46), the total fertility rate is seven children per woman. In Canada (birth rate: 11), it is about 1.5 children per woman.

Why are birth patterns so different in different parts of the world? And why does it appear that the countries that can least afford children have the highest birth rates? There are several answers to these questions. The fact that birth control is more available in developed countries is only part of the answer. The reasons for a high birth rate in poor countries include:

▸ More children are needed to help work in the fields to produce food for the family.

Figure 8
How do the total fertility rates of many African countries and Canada compare?

▸ More children are needed to support their parents in old age (since insurance programs and government support for the aged may not exist).

▸ More children are needed to replace the many who die before adulthood.

▸ Some countries have a culture or religion that favours large families.

Reducing the birth rate is one of the most difficult of all global challenges. It will only happen if people in high-birth-rate countries develop a more positive attitude toward having smaller families. Changes that might help them do so include:

▸ Improving the health of infants. People may decide to have fewer children if they can be confident that their children will survive into adulthood.

▸ Improving the status of women. In poor countries with high birth rates, women are often uneducated and have a low status in society. Whenever women in a country have been provided with a better education and other opportunities in life, the birth rate has always declined.

Figure 9
These young women are tending seedlings in Africa.

Staying Alive

In every country in the world, people try to prevent or delay death. Some countries have been more successful than others. Sadly, the death rates in the developing world are still much higher than those in the developed world.

A particular tragedy in the world's poorest countries is the high *infant mortality rate*. This is the number of infants per thousand births who die before the age of one. In poor countries the infant mortality rate is as high as 150, in spite of a recent reduction. In the last 20 years organizations such as UNICEF have made great efforts to reduce infant mortality by setting up immunization programmes. Another aid in reducing infant deaths is ORT (oral rehydration therapy)—the use of a sugar and salt solution to help children who are dehydrated.

In many developing countries, surviving past the age of one is only the first hurdle. A low standard of living and a lack of medical facilities keeps **life expectancy** low. In Malawi, the average person is expected to live to only 36. In Canada, life expectancy is 79. Japan has the highest life expectancy at 81.

life expectancy—the average number of years a person is likely to live. It depends on many factors, particularly the standard of living in a person's country of residence.

Future Population Patterns

In 1999 the world population passed the 6-billion mark. It entered the second millennium still growing at an annual rate of about 84 million, or 1.4 per cent. This growth rate may seem small, but it is enough to double the world's population in 50 years.

Figure 10 shows the predicted growth of world population with a medium birth rate. In your generation, world population is estimated to pass: 7 billion in 2010 and 8 billion in 2022.

Discover ☼ With Maps and Graphs

1. Using Figures 5 and 6 (pages 76 and 77), identify the continents or regions with the highest and lowest birth and death rates. Explain why continents differ, using information from pages 76 to 81.
2. Divide into small groups. Make a large group-copy of the graph of population growth in Figure 10 on page 82. Discuss where the

labels (a) to (d) should be placed on the graph. Add them to your graph using arrows.

(a) Death rates fall—advances in medicine and sanitation during the Industrial Revolution help to reduce infectious diseases.

(b) New medical programmes and drugs in the 20th century reduce death rates still further.

(c) Falling death rates bring huge increases of population in countries where the birth rate remains high.

(d) In industrialized European countries, lower birth rates bring population growth to near-zero levels.

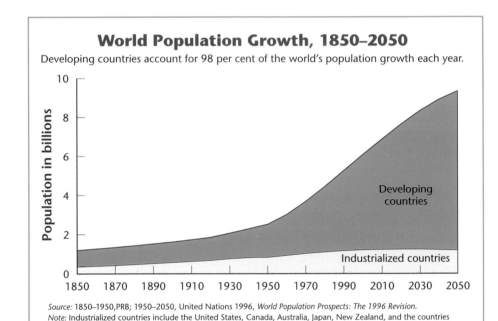

World Population Growth, 1850–2050

Developing countries account for 98 per cent of the world's population growth each year.

Source: 1850–1950, PRB; 1950–2050, United Nations 1996, *World Population Prospects: The 1996 Revision.*
Note: Industrialized countries include the United States, Canada, Australia, Japan, New Zealand, and the countries of Europe; all other countries are classified as developing.

Figure 10
World Population Growth in Developing and (Industrialized) Developed Countries

For up-to-date data on population growth, look up http://www. undp.org/popin/wdtrends/p98/p98.htm To follow the second-by-second growth in world population, look up http://www.census. gov/main/www/popclock.html Click on "Java version of World POP Clocks" and wait a few moments for the clock to appear.

Summary

In this chapter you have discovered that the world's population is growing quickly, especially in the world's poorest countries. You have learned how medical techniques and drugs have reduced the death rate. You have also seen that reducing the birth rate (or average family size) in poor countries remains one of the world's greatest challenges.

Reviewing your Discoveries

1. List the factors that encourage high birth rates in developing countries.
2. List the factors that encourage low birth rates in developed countries.
3. Describe the population patterns in a country whose population pyramid has a wide base. How are these patterns different from the patterns of a country whose population pyramid has a narrow base?
4. Explain why life expectancy in the world's richest countries can be twice as high as life expectancy in the world's poorest countries.

Using Your Discoveries

1. Do a group project on the population patterns of a country of your choice. Divide up the research among group members. Your project should include:
   a) a description of the country's age patterns, with a graph showing the percentages of the population in these age groupings: under 15 years, 15–64 years, 65 years and over
   b) statistics on and descriptions of: birth rate, death rate, infant mortality rate, life expectancy
   c) a graph showing your group prediction of the country's population growth from 2000 to 2050, based on b)
   d) an explanation of your prediction
   e) suggestions as to what the government's policy on population should be.

Chapter 6

Population Contrasts

Key terms

standard of living

gross national product
(GNP)

correlation

standard of living—the extent
to which people have the goods
and services they need and want.

In this chapter we focus on the different standards of living of
the people in different countries. The information and activities
will help you

▸ compare the characteristics of developed and developing
  countries
▸ show your understanding of the relationships between
  standard of living indicators
▸ analyze the educational skills needed around the world.

Measuring Living Standards

So far in this unit we have learned about the different kinds of
human patterns geographers study. They include patterns of settlement, land use, and population growth. Why are geographers
interested in these patterns? One important reason is to understand how they can affect people's **standards of living**.

Think for a few moments about your standard of living. How
well are your needs for food and water met? How are you able
to cope when you get sick? What things in your life have you
received or bought simply because you wanted them? What
opportunities do you have that young people in other countries
might not have?

Article 25 of the UN Charter of Human Rights (1948) states:
"Everyone has the right to a standard of living adequate for the
health and well-being of himself and of his family, including

food, clothing, housing and medical care and necessary social services…" Using this statement, we can begin to make a list of what standard of living includes. Each item in Figure 1 is an *indicator* of standard of living. There are ways in which we can measure each indicator. With such indicators, we can compare people's standards of living in different countries.

In this chapter we focus on the indicators of nutrition, health, education and literacy. First we shall look at some of the problems of using average income per person as an indicator.

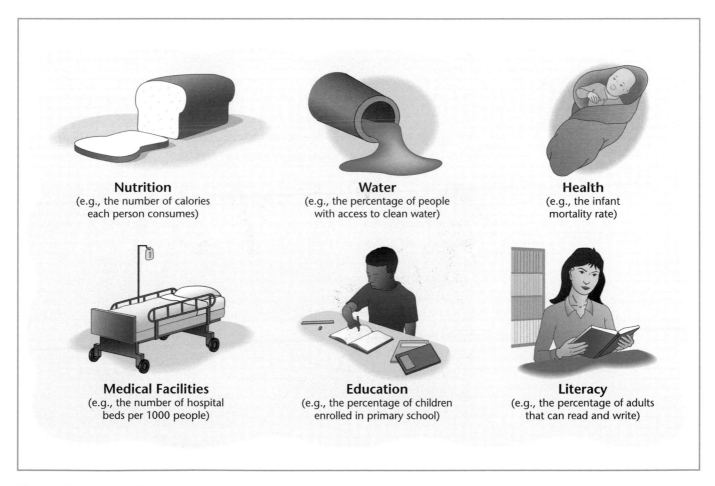

Nutrition
(e.g., the number of calories each person consumes)

Water
(e.g., the percentage of people with access to clean water)

Health
(e.g., the infant mortality rate)

Medical Facilities
(e.g., the number of hospital beds per 1000 people)

Education
(e.g., the percentage of children enrolled in primary school)

Literacy
(e.g., the percentage of adults that can read and write)

Figure 1
Standard of Living Indicators

gross national product
(GNP)—the sum of the value of
all goods and services produced
in a country in a year. It is often
measured in American dollars
(US$).

Income as an Indicator

If you want to compare how well off two people or groups of people are, you might look at their incomes. Economists have come up with a way of calculating the average income per person in a country as a basis for making comparisons. This calculation is based on the country's **gross national product (GNP)**. The GNP is divided by the country's population to give *GNP per person*. This term is the same as "average income per person" in the country.

When we describe countries as "rich" or "poor," we are influenced by differences in their populations' average incomes. But there is a problem with using this information to judge standards of living between countries: *money is used differently by people in different countries*. In Canada, we use money to pay for most of our needs and wants. In other countries, people grow their own crops to get their daily food requirements. They can meet a great deal of their needs while making and spending less than US$1 per day. Also, costs of living vary greatly between countries.

Figure 2
In countries such as Sudan, poor people can grow vegetables year-round instead of buying them.

WEB LINK

To find out more about standard of living indicators, look up
http://www.oecd.org/dac/Indicators/htm/maps.htm

 **For Yourself**

1. Why is "GNP per person" not an ideal way to measure standards of living?
2. Why do you think the following are not used as standard-of-living indicators?
   a) the types of houses people live in
   b) the types of clothing people wear
3. The organizer in Figure 3 divides up your standard of living into four categories of needs and wants.
   a) Make a copy of the organizer and fill it in.
   b) In which category would you put "TV set"? Explain your point of view.
   c) If your city provides TV sets for the elderly and invalids, would you consider this an action to meet a need or a want? Explain.

Things that I *must* have e.g. adequate food & water	Things I *want* to have e.g. mountain bike	Things I *want* to do e.g. go to hockey games	Services I *need* in life e.g. doctor

Figure 3
Organizer for Question 3

4. In small groups, make another organizer with these five headings: Food, Clothing, Education, Health, Leisure. Discuss the things under each heading that you think all the people in the world have a right to have. Prepare a written summary of the conclusions you reached as a group. Include a description of any disagreements you had and explain why you think the disagreements arose.

Nutrition

Food provides the human body with energy from three sources: *starch* (carbohydrates), *fats*, and *proteins*. Proteins have the added function of maintaining the health of the cells in our bodies. We also need small amounts of certain vitamins and minerals. Most of us get all these needed elements by eating a variety of foods each day.

People with a poor standard of living suffer from *malnutrition*. Although this term really means "bad nutrition," it usually refers to a lack of food. Figure 4 shows how malnutrition leads to the problem of underweight children in the developing world.

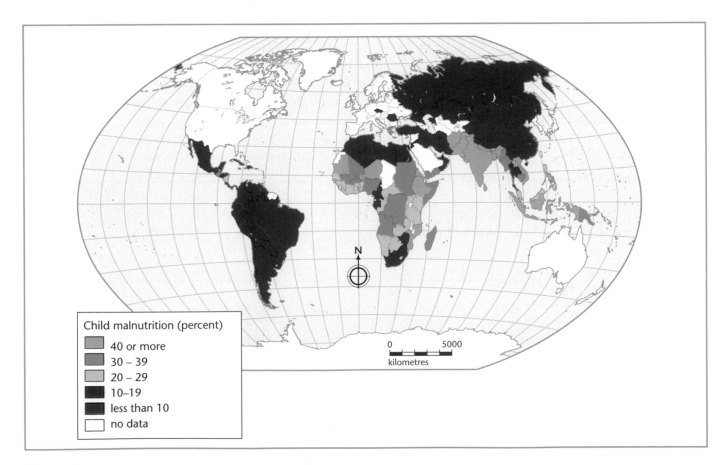

Figure 4
The Percentage of Children Who Are Malnourished in Developing Countries

Improving Nutrition

In 1996, the Food and Agriculture Organization (FAO) of the United Nations met in Rome, Italy. There, it was reported that globally "190 million children are underweight, 230 million children are stunted and 50 million children are wasted (severe malnutrition)." A resolution was passed to reduce the number of seriously undernourished people in the world by half—from more than 800 million to 400 million—by the year 2015.

To solve the problems of malnourishment, changes need to be made in how food is grown, marketed, and stored. Right now, much of the world's farming land is used to produce food (mainly grass) for animals so that the animals can provide food for humans. In this type of farming, people receive only about one tenth of the energy contained in the plants that the animals eat. Growing food crops instead of animal grasses would feed many more people. This solution is unlikely, however, because it would reduce the choices people in richer countries have to buy their favourite foods.

Food storage is also a major problem, especially in the developing world. It has been said that it would take a train 4000 km long to transport the amount of grain eaten by India's rats in a year.

Solutions to these problems include the following:

▸ using land to grow cereals for people instead of food for animals
▸ improving and extending irrigation systems
▸ using new crop varieties that produce higher yields
▸ paying farmers better prices for their crops to encourage them to grow more
▸ teaching people how to store food properly.

WEB LINK

To find out more about farming in developing countries and the importance of women farmers, look up
http://www.fao.org/gender/en/nutr5-e.htm

Discover  With Maps and Graphs

1. Look at the world map in Figure 4 on page 88. Which countries have the largest proportion of the world's underweight children?
2. Look at the graph in Figure 5.
   a) Which region is predicted to have the most undernourished people by the year 2010? Which region is likely to improve most?
   b) Figure 6 shows how hard a woman in this region must work to produce food for her family. In what ways is the life of this woman different from the life of an adult woman in Canada?
3. Why should most of the money from relief agencies for countries with malnourished people go toward providing long-term development assistance rather than short-term food aid?

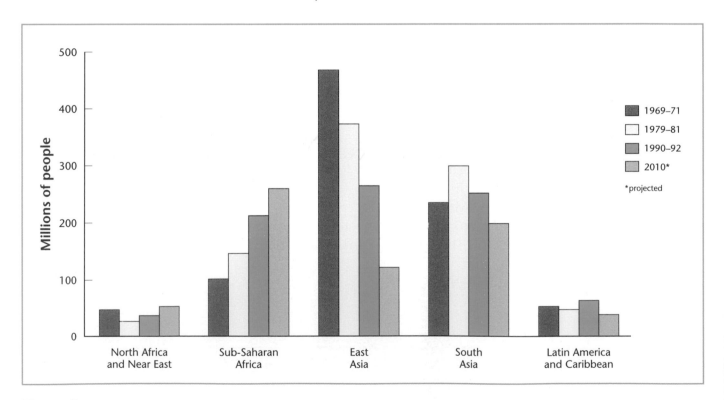

Figure 5
Estimated Number of Malnourished People in Different Regions of the Developing World, with Projections for 2010

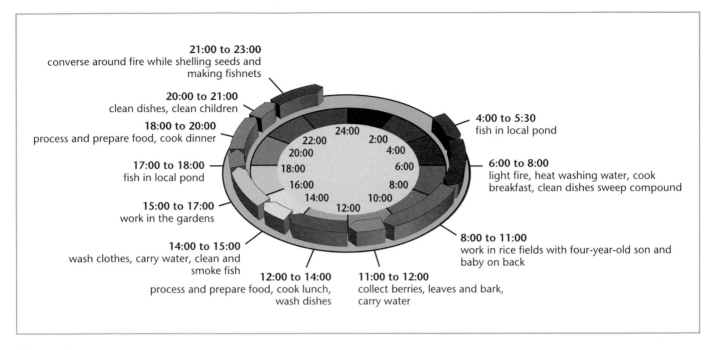

Figure 6
One Woman's Day in Sierra Leone

Case Study   *Health Hazards in the Developing World*

Medical discoveries such as the smallpox vaccine, penicillin and other antibiotics have dramatically reduced the death rate in both developed and developing regions. Developing regions have also greatly benefited from medical help provided by non-governmental organizations (NGOs), the World Health Organization (WHO), and the United Nations Children's Fund (UNICEF). Yet hundreds of millions of people in the developing world still suffer from diseases unknown to many in the developed world. The main problem is the humid, tropical climate of many developing countries, in which bacteria and insects that carry disease breed quickly. *Malaria* and *bilharzia* are two examples.

WEB LINK   To find out more about UNICEF, look up
http://www.unicef.org/

Malaria

Malaria affects 40 per cent of the world's population in 100 countries, mainly in the tropics. It causes fever, shivering and aching joints. It is often fatal, especially when combined with the effects of other diseases. Every year, about one million children under the age of five die from malaria. Countries in tropical Africa account for more than 90 per cent of malaria cases and the great majority of malaria deaths.

The female *anopheles mosquito* carries the tiny parasites that cause malaria. These parasites infect the blood taken by a mosquito when it bites a person. This infected blood is passed to the next person from whom the mosquito takes blood and destroys the person's red blood cells.

Malaria has been getting worse, mainly because most varieties of the parasite have now developed a resistance to the drugs used to treat them. The main solutions are to try to control the breeding of mosquitoes by draining ponds and other areas of stagnant water where mosquitoes breed, and by spraying chemicals. Scientists are also working on the vaccine to immunize people so that they don't catch malaria when bitten by infected mosquitoes.

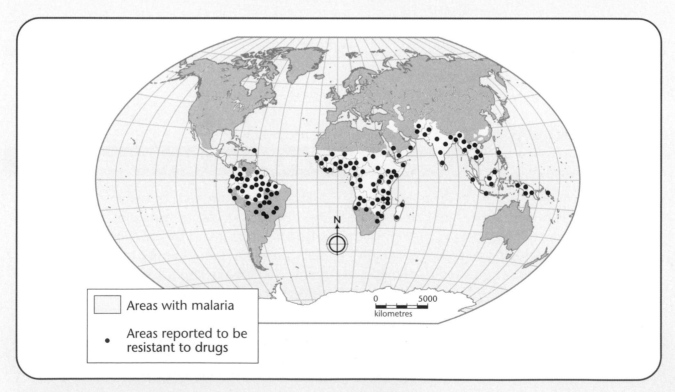

Areas with malaria

Areas reported to be resistant to drugs

Figure 7
Areas of the World with Malaria and Areas Where Resistance to Drugs is a Problem

Bilharzia

This disease (also called Schistosomiasis) is related to water as well. It may be thought of as a cycle, beginning when parasites enter the body through the feet or other body parts exposed to water. The parasites work their way through the bloodstream to the bladder and the liver and lay eggs. The sharp eggs puncture the bladder, are excreted out of the body and re-infect the water.

The next part of the cycle takes place if the eggs hatch and the tiny parasites find a fresh water snail. They enter the snail, divide into greater numbers and then leave the snail, ready to enter the human body.

Increased irrigation has led to the spread of bilharzia. Two hundred million people in 74 countries are now infected. They can be helped by the following:

- drugs that kill the parasites in people
- chemicals that control water snails
- better sanitation facilities
- learning better hygiene practices.

For Yourself

1. Explain why the death rate has fallen so quickly in developing countries.
2. Why would vaccination be the most suitable method of controlling malaria? Refer to Figure 7 in your answer.
3. Which of the possible methods of controlling bilharzia do you think is best? Explain your choice.

Literacy and Education

Literacy is the ability to read and write. Without this ability, a person is at a serious disadvantage in the modern world. For example, an illiterate person cannot read instructions or fill in a job application. Clearly, a country's **literacy rate** must be as high as possible if its people are to enjoy a better standard of living.

Canada, like other developed nations, has a high literacy rate. The statistics show a rate of over 95 per cent. But surveys have shown that the number of Canadians who cannot *adequately*

literacy rate—the percentage of adults (people over the age of 15) who can read and write.

read or write is well over 5 per cent. The literacy skills needed in Canada today involve reading at high levels (for example, to understand computer manuals or information on other types of technology).

The least developed countries of the world generally have the lowest levels of literacy. People in these countries need to be educated, not only in literacy, but also in *numeracy* (counting and using numbers) and in life skills. As Figure 8 shows, educating girls in particular can have a dramatic effect on the quality of life in developing countries.

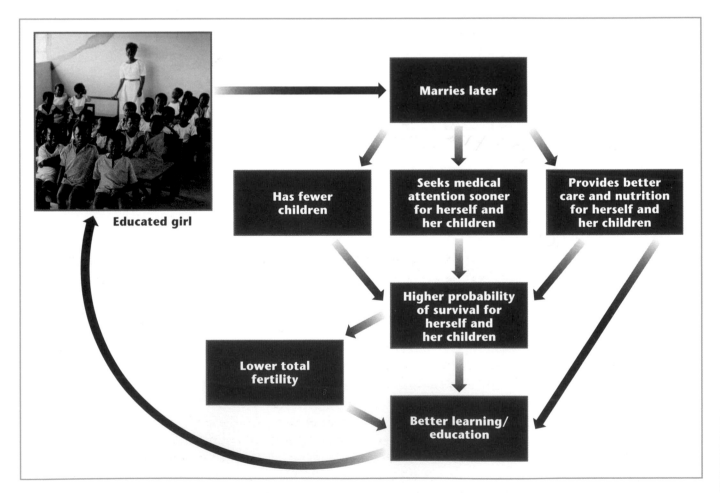

Figure 8
Use this diagram to explain how education improves the quality of life in developing countries.

To improve educational levels in developing countries, the following obstacles must be overcome:

▸ Birth rates must be lowered so that the number of children is not greater than the number of places in schools.

▸ The economies of poor countries must be made stronger so governments do not cut education budgets to help pay off debts.

▸ Transportation systems, including school buses, must be improved so that children in rural areas can get to school.

The figures below show that some countries have a long way to go in achieving these goals.

Figure 9
Graph (a) shows the percentage of all school-age children who enrol in school. Graph (b) shows the percentage of those children who reach Grade 5.

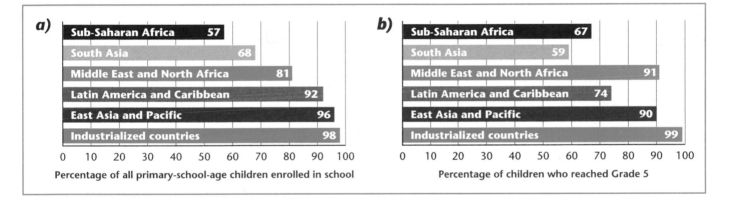

a)

Sub-Saharan Africa	57
South Asia	68
Middle East and North Africa	81
Latin America and Caribbean	92
East Asia and Pacific	96
Industrialized countries	98

Percentage of all primary-school-age children enrolled in school

b)

Sub-Saharan Africa	67
South Asia	59
Middle East and North Africa	91
Latin America and Caribbean	74
East Asia and Pacific	90
Industrialized countries	99

Percentage of children who reached Grade 5

Case Study

The BRAC Project in Bangladesh

Bangladesh is one of the poorest countries in South Asia. It has a dense population and a low standard of living as measured by nutrition and health. Many organizations, some funded by the Canadian International Development Agency (CIDA), have been working to help the people of Bangladesh.

The Bangladesh Rural Advancement Committee (BRAC) was set up in 1985 to provide health facilities and credit to farmers. One of BRAC's aims was to provide basic literacy and numeracy to 8- to 10-year-old children. Small schools of about 30 children (usually about 20 girls and 10 boys) have been set up to meet for three hours a day, 268 days a year. Educated members of the local community help teach the children, and the school calendar can be adapted to fit local events such as harvests. By 1998, 34 000 BRAC schools were operating to improve children's levels of education. This type of project has helped to reduce the country's birth rate from 48 per thousand in 1970 to 27 per thousand in 1999.

Discover ☀ For Yourself

1. Explain how the work done by BRAC could result in a fall in Bangladesh's birth rate.
2. Divide into small groups to compare your education with education in the developing world. Discuss these questions:
   a) What skills needed by a Canadian student might be different from those needed by students in BRAC schools? Give three examples. Why might these differences exist?
   b) What types of equipment and facilities in your classroom do not appear in the African classroom in the photograph (Figure 8)?
   c) If you wanted to become a teacher in Bangladesh, how might your training be different from the training to become a teacher in Canada?
3. In your groups, discuss how what you learn today in school is different from what your grandparents learned. Make a list of the types of skills you need today that were not needed in the past.

Figure 10
A Grade 8 Classroom in Canada

O⊓

correlation—a relationship or pattern between two factors that is fairly predictable. The factors compared in correlations are usually described by sets of numbers.

How Indicators are Related

Our study of education in the developing world shows a relationship between female literacy and birth rates. The higher the rate of female literacy, the lower the birth rate. Geographers are always looking to find such **correlations** to help them understand the human world. Discovering correlations makes it easier to take action to solve problems. For example, based on the correlation we've just seen, providing better education for girls in developing countries is an action governments could take if they wanted to lower the country's birth rate.

When geographers find that an increase in one factor goes along with an increase in another factor, they have discovered a *positive* correlation. An example of a positive correlation in everyday life involves the factors of *height* and *shoe size*. Taller people tend to wear *larger* shoes. A *negative correlation* is the opposite case. For example, the *more* times you use a ballpoint pen, the *less* amount of ink it has.

Discover With Graphs

1. Figure 11 on page 98 includes information on a number of standard-of-living indicators in a number of countries.

   a) Draw a horizontal and vertical axis for a graph. Make your horizontal axis 10 cm long and label it "Female Literacy Rate (%)." Make your vertical axis 6 cm high and label it "Birth Rate."

   b) Make a small mark at every centimetre along the horizontal axis. Label the small marks with multiples of 10 (e.g., 10, 20, 30,...,100).

   c) Make a small mark at every centimetre along the vertical axis. Label the small marks with birth rates in multiples of 10 (e.g., 10, 20, 30,...,60).

   d) For each country in Figure 10, mark a dot that is directly above its literacy rate and directly across from its birth rate. You will end up with a *scattergram*—a graph with scattered points.

   e) Draw a straight line on the graph showing the general trend, or best fit, of the dots. (Note that the line will not join the dots; some dots will be above the line and others below it.)

   f) What is the significance of the fact that the line slopes down from left to right? In other words, does this graph show a *positive correlation* or a *negative correlation* between the two items?

   g) Explain in your own words what this tells you about the relationship between female literacy and birth rates?

MATH
LINK

Turn to page 263 for more practice making scattergrams.

Country	Literacy Rates in 1995 (%)			Birth Rate	Life Expectancy
	Male	Female	Total		
Afghanistan	47	15	32	43	46
Algeria	74	49	62	30	68
Brazil	83	83	83	21	67
China	90	73	82	16	71
Ethiopia	46	25	35	46	42
India	66	38	52	28	60
Italy	99	98	98	9	78
Niger	21	7	14	54	41
Peru	95	83	89	28	68
Sri Lanka	93	87	90	19	72

Figure 11
Use this data on standards of living in different countries to answer Question 1 on page 97.

2. Draw another scattergram with a 10 cm-long horizontal axis and an 8 cm-high vertical axis.
   a) Label the horizontal axis "Total Literacy Rate (%)." Label the vertical axis "Life Expectancy in Years."
   b) Repeat steps b) to e) from Question 1 (except this time each dot will be directly above the country's total literacy rate and directly across from its life expectancy).
   c) What type of relationship does this graph show? (Hint: in a positive correlation, one item rises as the other rises.)

Summary

In this chapter you have discovered some ways that geographers measure standards of living throughout the world. You have learned that certain standard-of-living indicators are correlated. You have also reflected on the challenges faced by developing countries in providing nutrition, health, and education to their citizens.

Reviewing Your Discoveries

1. Explain what is meant by "standard of living".
2. List three ways in which your standard of living might be higher than that of a person of your age in a developing nation.
3. Explain why there are more diseases in developing countries than in developed countries.
4. Give three reasons why it is difficult to educate children in developing countries.

Using Your Discoveries

1. Do a group project on the standard of living of a country you think is much less developed than Canada. Divide up the research among the group members. Your project should include:
   a) how the country measures up with respect to four different indicators
   b) why the country has a low standard of living
   c) what could be done to improve the standard of living
   d) how Canada could contribute to this improvement.
2. Some people think that food and development aid should not be given to developing countries because it might encourage population growth.
   a) Take a stand for or against this viewpoint. Think of reasons for your stand.
   b) Debate your stand with someone with an opposing viewpoint. Use some kind of media to present your opposing stands (for example, a live debate, a video, a pamphlet).

Figure 12
These people are awaiting the distribution of food aid at Bixen Duule refugee camp in Northwest Somalia.

Geography Workshop

This unit has presented five topics that geographers study to find out about human patterns:

S settlement types
L land use in settlements
U urbanization—the move to cities
G growth in the world's population
S standards of living in different world regions

In this culminating activity, you will apply what you know about these topics to plan a city of about 100 000 people.

The Situation

In a small group, draw a map of a piece of landscape similar to the one shown on the opposite page. Choose a site for your city. The site should be suitable for one of the following types of cities:

▸ a port city with oil refining and grain-milling industries
▸ a retirement or tourist city
▸ a developing-world city coping with many immigrants
▸ an industrial city with a large car assembly plant

The Plan

Make a colour-coded land use map that shows your plan for the city you chose. The map should show the following:

▸ Central business district
▸ Railways and main roads
▸ Main shopping centres
▸ Industries
▸ Recreational areas
▸ Schools and other public buildings

▸ Residential areas, as follows:
a) low cost housing (shantytown if you chose the developing-world city)
b) medium cost housing
c) high cost housing

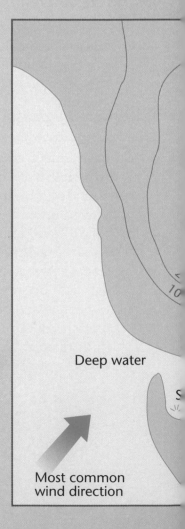

Deep water

Most common
wind direction

The Profile

Prepare to present your city plan to a government committee. Your presentation should include a profile of the city's population with four sections:

1. A sketch of a population pyramid showing the ages and sexes of the city residents
2. A bar or circle graph showing the possible occupations of the city residents
3. An explanation of how the city plan suits the city's physical and human characteristics
4. Predictions about these features of the city in 50 years' time:
   a) population characteristics
   b) employment patterns
   c) social problems
   d) environmental problems

Reflecting On Your Work

Describe how you used your understanding of human patterns to help you create the following: a suitable city plan, a realistic city profile, and reasonable predictions about the town's future.

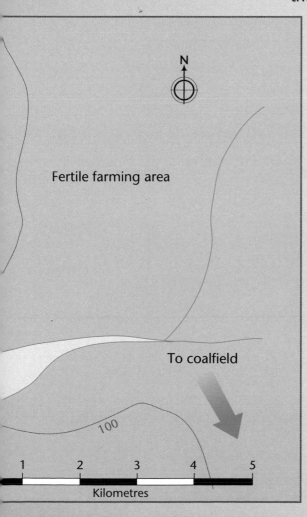

N

Fertile farming area

To coalfield

100

1   2   3   4   5

Kilometres

Unit Two

Discovering Economic Systems

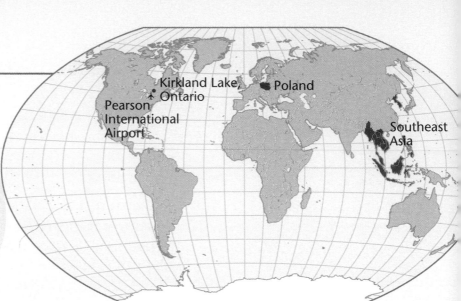

In Unit One, you learned that your standard of living is linked to your needs and wants. If you can meet your needs and get the things you want, you have a high standard of living. People have created systems to make these goods and services available to you. These systems are the focus of economics.

This unit will help you discover the different types of economic systems in the world. You will learn about the specific features of these systems—what they produce, how they produce, and how the benefits are distributed. You will see how these features combine to make up Canada's economy. You will also look at how Canada's economy is linked to economic activity in other countries. While you do so, you will "visit" places and regions around the world, including those highlighted in the world map on this page.

Chapter 1

Economics and You

Key terms

economics
scarcity
entrepreneur

economics—the study of the production, movement, distribution, marketing and consumption of goods and services. Specifically, economics deals with *which* goods and services we produce, *how* we produce them, and how we *distribute* them.

In this chapter we focus on how economics is a part of our lives. The information and activities will help you

▸ show your awareness of the main elements of an economic system
▸ show your understanding of the influence of entrepreneurs
▸ compare how countries in the global community consume resources.

Economic Systems

Every day your life is somehow involved with **economics.** This involvement includes buying goods (products such as clothing or food). It also includes using services (the things people get paid to do for other people, such as giving piano lessons, a haircut, or a ride on a bus). Economics occurs around you, too, as people buy, sell, and exchange larger goods such as houses, and services such as sewage treatment.

Goods and services are important to us because they satisfy our needs and wants. In order to produce goods and services, we need *resources* such as land, labour, and factories.

Unfortunately, we can never have enough resources to provide for *all* our needs and wants. In other words, we face the problem of scarcity. This problem forces us to choose certain needs or wants over others. We develop economic systems to help us make these choices.

You deal with scarcity in one way or another every day of your life. For example, your time is scarce, so you must choose

Figure 1

Describe the resources shown in these photographs. How is the idea of **scarcity** connected to these resources and to the need for economic systems?

which activities you will spend time on and how long you will spend on each of them. When you spend a resource such as time or money on one thing, you give up the opportunity to spend it on another thing.

Just as you as an individual make these choices about your resources, a society also makes choices about its resources (also called *factors of production*; we will look at these factors of production in detail in Chapter 3). The decision-making process is summarized in Figure 2.

scarcity—the limits in the amount of resources we have. While our resources are scarce (limited), our wants are great (unlimited).

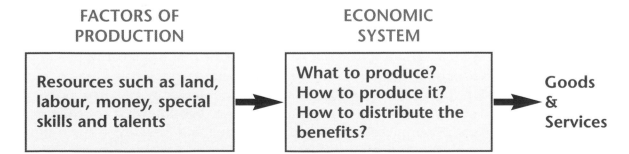

FACTORS OF PRODUCTION

Resources such as land, labour, money, special skills and talents

ECONOMIC SYSTEM

What to produce?
How to produce it?
How to distribute the benefits?

Goods
&
Services

Figure 2

Economic systems answer the three basic questions of economics.

Discover For Yourself

1. Classify each of the following items as a *good* or a *service*:
   a) carpet
   b) dry cleaners
   c) library
   d) book
   e) school
   f) restaurant
   g) box of cereal
2. List 10 different items your family spends money on in a typical month. At least three of them should be services.
3. Show the list you created for Question 2 to someone such as a grandparent, or try to find someone who is at least 60 years old.
   a) Ask the person how the items on your list compare to the items they would have spent money on 50 years ago.
   b) Suggest two changes that have taken place in the last 50 years that would explain any differences.

Turn to page 129 for examples of successful Canadian entrepreneurs.

entrepreneur—a person who takes a risk to start and run a business.

Participating in the Economy

When we buy goods and services, we become *consumers*. As consumers we interact with people in many other economic roles. *Wholesalers* and *retailers* sell goods and services. *Advertisers* persuade us to buy goods and services. *Market researchers* find out what goods and services we want. *Researchers and developers* improve existing goods and services and create new ones.

At the beginning of it all are the *producers*, who make or provide goods and services. A special kind of producer in our economic system is the **entrepreneur**. Entrepreneurs often introduce new goods or services to the marketplace. They must be especially skilled in abilities such as defining what consumers want, being their own boss, motivating and inspiring others, and taking reasonable risks. In addition to these qualities, successful entrepreneurs are often lucky enough to be "in the right place at the right time."

Case Study

McDonald's Restaurants

McDonald's and many other restaurants are hugely successful because they meet the demand for fast food. This demand first grew in the late 1940s and early 1950s, when the lifestyles of North Americans became much busier than ever before. The time available for preparing and cleaning up after meals started to shrink dramatically.

The idea for "fast food" as a solution to this problem was first worked out by the McDonald brothers, who had a hamburger restaurant in San Bernardino, California in the 1940s. Their drive-in restaurant brought in US$200 000 yearly, but they wanted to do better. They realized that the way to lure in customers was to serve them meals more quickly and at a lower price. After much thinking and experimenting they decided to

- *speed up food preparation* with a reorganized kitchen, assembly-line methods and a reduced menu of 9 items from 25;
- *speed up service* with a self-serve counter to replace bellhop service to cars; and
- *lower the cost* of a hamburger to 15 cents from 30 cents.

The McDonald brothers reopened their business in December, 1948. By 1955 they had yearly sales of US$350 000. Many people wanted to operate similar restaurants. The brothers sold a **franchise** to Neil Fox, who established a restaurant in Phoenix, Arizona.

Figure 3
Ray Kroc stands in front of an early McDonald's Restaurant.

An entrepreneur named Ray Kroc was so impressed by the McDonald's operation that he became the sole agent for selling franchises. In 1961, he was willing to take the reasonable risk of buying out the McDonald brothers. Four years later, Kroc began selling shares of the business to raise money to expand the chain. Since he was short of money at the time, he paid many employees in shares. One hundred shares purchased in 1965 at $22.50 each would be worth $1.8 million today. These employees became very wealthy when they eventually sold their shares.

franchise—the right to sell a particular good or service. Each franchise operation must follow guidelines to make it similar to other franchises owned by the same company.

In 1967, the first McDonald's was opened in Canada. The early 1970s saw expansion into the Caribbean Islands, Japan, Germany, Australia, France, and Great Britain. In the 1990s, openings in Moscow (Russia), Beijing (China), and Tel Aviv (Israel) were big news. By 1998, McDonald's had 23 000 restaurants worldwide, 13 000 of which were spread across the United States. Sales in 1998 were almost $12.5 billion.

McDonald's Franchises

Ray Kroc sold his first McDonald's franchises for $950 each. A franchise permitted use of a McDonald's building and the names and recipes for different McDonald's food products. Even back in the 1950s, the buildings had many of the design features we recognize today. Red and white tiles adorned buildings with sloping roofs and a "Golden Arches" design on the side. Franchise owners paid 1.9 per cent of their sales to Kroc who, in turn, paid 0.5 per cent of sales to the McDonald brothers.

A franchise today still contains the right to use McDonald's buildings, food recipes, logos, and so on. But it now costs US$300 000, payable to the head office in the United States. Many restaurants take in between $1-2 million in sales each year. About 16 per cent of the value of annual sales is sent back to the United States each year. One quarter of this is used for company-wide advertising.

The franchise owner is expected to spend some of the profits in support of neighbourhood activities and local charities. The franchise's meat, buns and other items are bought from local suppliers whenever possible. These guidelines are meant to give McDonald's a positive image in the community. McDonald's also provides jobs for nearby residents, particularly for students who work part time.

 **For Yourself**

collateral—something owned by a person taking out a loan that can be seized if the person does not repay the loan.

1. Why is the fast-food industry so profitable?
2. Ray Kroc bought out the McDonald brothers for US$2.7 million. He took out a bank loan for this amount, using the value of the buildings he owned as **collateral**. With interest, it cost him $14 million to repay the loan. Divide into small groups to discuss these questions:

 WEB LINK

To find out more about McDonald's, look up
http://www.mcdonalds.com/corporate/index.html and
http://www.mcdonalds.com/surftheworld/index.html

a) What are the advantages of borrowing large sums of money for a business? What are the disadvantages?

b) How would you guess the McDonald's operation might have developed differently if Kroc had waited until he had earned the money himself to buy out the brothers?

c) What other method of raising money did Kroc use?

3. Why is being an entrepreneur risky? Include examples from Ray Kroc's experience in your explanation.

4. As the McDonald's chain has expanded, it has added new features to its operation: *breakfast* (introduced in Hawaii, 1970); *Playland* (introduced in California, 1971); *drive-thru* (introduced in Arizona, 1975).

a) Explain why each of these new features increased sales.

b) Which feature do you think increases sales the most? Why?

c) Suggest another feature that you think McDonald's should add. Explain why you think it would increase sales.

All Consumers Are Not Equal

Even though goods or services such as McDonald's restaurants can be found worldwide, not everyone has the same opportunity to consume them. In many parts of the world, consumers are not able to buy as much as most Canadians. They may work as hard or harder than many Canadians, but their pay is often very much lower. As a result, they cannot buy the same kinds of goods and services that we do.

One way of measuring the wealth of people in different countries is GNP per person. Although this measure is not the most reliable indicator of standard of living, it is a helpful way to compare how much people can buy. This in turn tells us how much they are a part of world economic systems. Canada has one of the highest GNPs per person, at US$19 640 in 1997, while Ethiopia and The Democratic Republic of the Congo have the world's lowest, at US$110 in 1997. Figure 4 on page 110 shows how the rest of the world compares to these extremes.

Turn to page 86 to review GNP per person.

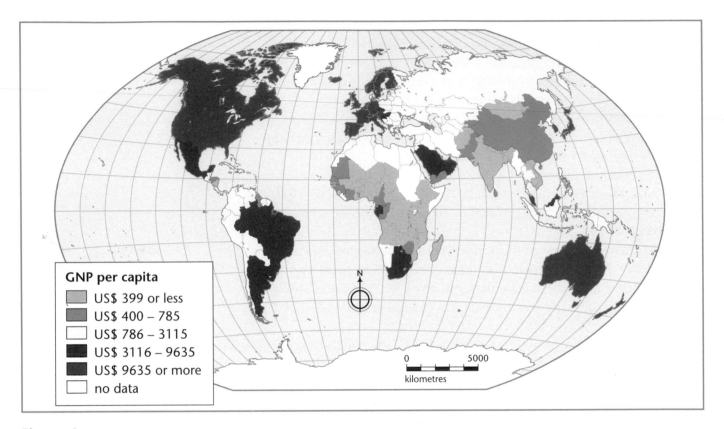

GNP per capita

	US$ 399 or less
	US$ 400 – 785
	US$ 786 – 3115
	US$ 3116 – 9635
	US$ 9635 or more
	no data

Figure 4
GNP Per Person in Countries
Around the World

Discover **For Yourself**

1. Rank the countries in Figure 5 by GNP per person. Where would Canada rank if it were included in this list?

2. Draw a vertical bar graph to display your ranking. Finish your graph by colouring and labelling the bars and axes, and adding an appropriate title.

3. List three countries from Figure 5 where you believe it would be most profitable to sell cars. Give reasons for your choices. What other information besides GNP per person might help you decide in which countries cars could be profitably sold?

4. List two countries from Figure 5 where you would not choose to set up a fitness club for local residents. Explain why not.

5. Why does a low GNP per person make it difficult for a nation to improve people's standard of living?

Country	GNP/person (US$)
Australia	20 650
Bolivia	970
Brazil	4 790
Ethiopia	110
Greece	11 640
India	370
Israel	16 180
Nigeria	280
United Kingdom (UK)	20 870
United States	29 080

Figure 5
Gross National Product per Person for Selected Countries, 1997

Economics and the Environment

All of the goods and services we buy use up some of the Earth's resources. For example, cars consume one third of the world's production of oil. Animals used for food must be fed from grains grown on pasture land. Pastures are created by clearing forests. Often, the cardboard packaging that our goods come in has been supplied by rain forest trees.

Each of us is using twice as much copper, energy, meat, steel and wood as was used in 1950. We are also using four times as many cars, five times as much plastic and seven times as much aluminum. The majority of these increases have occurred in the richer nations of North America, Western Europe, Japan and Australasia, which make up one fifth of the world's population. These nations have released two thirds of the greenhouse gases, three quarters of the gases that lead to acid rain, and 90 per cent of the chemicals that have damaged the protective ozone layer. Just imagine what would happen to the Earth if *all* its people used as many resources and produced as many pollutants as those in richer nations!

Figure 6
Share of Population and Resource Consumption

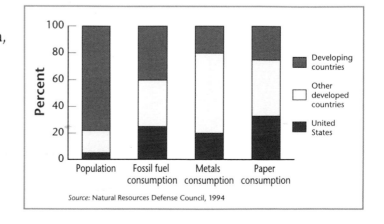

Source: Natural Resources Defense Council, 1994

 **Discover** With Graphs

1. According to Figure 6 on page 111, the United States and other developed countries make up 22 per cent of the population.
   a) What percentage of the world's fossil fuel supply do they consume?
   b) What percentage of metals do they consume?
   c) What percentage of paper do they consume?
2. What does Figure 7 tell you about cars and consumers in the developed versus the developing world? How does the pattern you see relate to one of the patterns shown in Figure 6?

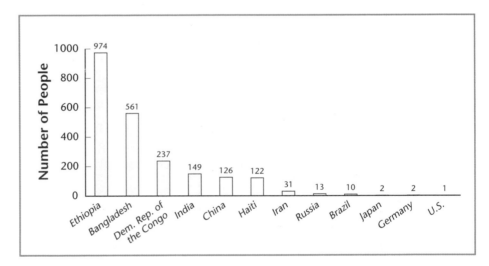

Figure 7
Number of People per Vehicle in Selected Countries, 1994

Summary

In this chapter you have learned about the ways people participate in economics. You have seen that entrepreneurs play a special and interesting role in providing goods and services. You have also discovered that different parts of the world have different levels of consumption and resource use.

Reviewing Your Discoveries

1. Give an example of
   a) a good or service that satisfies a need or want.
   b) a scarce or limited resource used in the production of the good or service.
   c) any environmental problems that result from the good or service.
2. Describe two problems caused by increased consumption.

Using Your Discoveries

1. You may have talents and interests that could lead to the development of your own business. For example, if you play a musical instrument, you might form a band that hires itself out for gigs. If you love to watch wild birds, you might start your own shop supplying the needs of people with backyard feeders.

   a) Form small groups. Make a list of the talents and interests of each group member.
   b) Pool your ideas on how these talents and interests could be developed into businesses. Choose one of the business possibilities.
   c) List the kinds of equipment, buildings, supplies, employees and other resources that you would need to start and run the business. Explain why you would need each of them. Mark any that you might have already.
   d) Suggest possible sources of money to buy the items you need.
   e) Explain how you would repay those who had financed your business.
   f) Describe and give reasons for the location that you would like to use.
   g) Design a logo/slogan for your business.
   h) Give an example of two kinds of advertising that you would use to attract customers.

Chapter 2

Canada's Economic Evolution

Key terms

O━

primary industries
secondary industries
tertiary industries

In this chapter we focus on the development of Canada's economic system. The information and activities will help you

▸ identify and give examples of three major types of industries
▸ describe how these industries have changed in importance
▸ identify patterns in the location of industries.

A Resource-Rich Country

Canada is well-known around the world for its abundant natural resources. These resources were first used by Aboriginal peoples. They hunted wild animals, caught fish and gathered foods such as berries, fruits, roots and seeds. In addition to supplying food, forests provided Aboriginal peoples with wood to help them build shelters. They used clay to make pots. Where the soil and climate were suitable, they were involved in agricultural activities such as growing corn, beans, and squash on land that they cleared.

When French and British explorers first reached the North American continent, they sent back word of the abundant resources they found. The French and British governments became especially interested in Canada's fish and fur resources. By the 17th century, fishing settlements had been established along the eastern coast, and fur trading had become important across the land. The fish were salted and sent back to Europe. Furs were traded from Aboriginal trappers and also sent back to Europe.

Starting in the mid-17th century, European settlers cleared land suitable for farming. The first areas to be cleared were along the eastern coast and the shores of the St. Lawrence River. Settlement and farming then spread to the shores of the lower Great Lakes

Figure 1
In the 1600s, Canada's economy included the traditional economic activities of Aboriginal peoples as well as fishing and fur trading. The fish and furs were exported to Europe.

and eventually to the Prairies. The timber from cleared forests and food products from farming were sent to Britain, which did not have enough wood and grain to meet its needs. Some of these resources were also sold to the United States. Figure 2 shows the movements of goods resulting from these economic activities.

Once railways were built across the Canadian Shield, valuable minerals could be transported to southern markets. The 1920s saw the opening up of the great metal belt from Sudbury to Cobalt and Porcupine. In other areas, railways or roads were built especially to transport the minerals south.

Turn to pages 5–14 to review settlement patterns in Canada.

Figure 2
Trade in Canada Between the Mid-18th and 19th Centuries

Discover With Maps

1. Form small groups to study the map in Figure 2. Discuss these questions:
   a) What three possible routes might furs originating in Western Canada have taken on their journey to Britain? (You will need to look at a map of North America for one of the routes.)
   b) How did the movement of furs out of Hudson Bay affect settlement on the coast of the bay? Name the settlements that resulted.
   c) What goods other than furs were exported from Canada in the 18th and 19th centuries? Where did they go? What were the benefits of this trade?
   d) What goods were imported into Canada? Where did they come from? What were the benefits of this trade?
   e) What role do transportation systems seem to play in the development and trade of natural resources?

The Three Sectors of Industry

In 1881, 48 per cent of Canadian workers made their living by farming. Most of the remainder worked in forestry, fishing, and mining. Economists group all of these activities together as resource industries, also called **primary industries**.

As machinery and technology developed, Canadian workers were able to use the products from primary industries and make them into other items. Some workers were actually forced into these new types of jobs, as machines replaced many labourers on farms and fishing boats, and in forests and mines. Soon, manufacturing industries (also called **secondary industries**) became established in villages, towns, and cities. These manufacturing industries made everything from prepared foods to furniture, machines, and clothing. By 1911, 25 per cent of Canadian workers were employed in secondary industries, while 39 per cent worked in the primary sector.

The development of secondary, or manufacturing, industries is important to a country's economy because these industries make the country's raw materials more valuable. To see how this is so, think of a loaf of bread.

primary industries— industries that harvest raw materials or natural resources (e.g., agriculture, forestry, fishing, mining).

secondary industries— industries that convert raw materials into finished products.

The production of a loaf of bread requires raw materials (the ingredients) *as well as* human labour, electricity and capital resources (for example, the oven). It is these other factors that actually turn the ingredients into bread. In doing so, they make the ingredients more valuable than they first were. The amount added to the product's value during the stages of production is called the "value-added."

Because of the growth of secondary industries and the value-added factor, prosperity grew, and living standards increased further. People were willing to pay for services such as schools, hospitals, police, and haircuts. These are service industries (also called **tertiary industries**) that do not make a product but improve the quality of life in some way. These services are paid for either by the person using the service or through taxes. By 1911, 33 per cent of Canadian workers were employed in this sector of the economy.

tertiary industries— industries that provide services (e.g., banking, retailing, education).

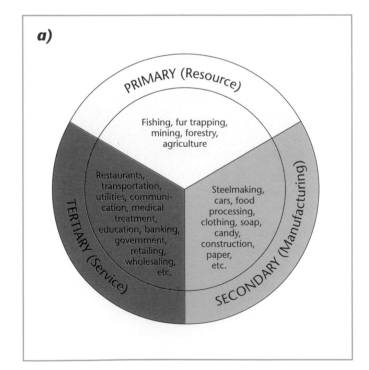

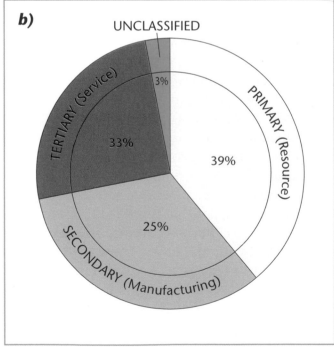

Figure 3
a) The Three Sectors of Industry b) The Structure of the Canadian Labour Force in 1911

Discover For Yourself

1. Figure 4 shows the numbers of Canadian workers in various industries in 1997.

   a) Group the industries into Primary, Secondary, Tertiary, and Unclassified. (Unclassified means that these industries do not fit easily into one category.)

   b) Calculate the percentage of the workforce in each of your four divisions. If your calculations are correct, their total will be 100.

   c) Make a circle graph to show your percentages. Your graph should resemble Figure 3(b). Colour or shade in and label the sectors as in Figure 3(b). Give your graph a title.

2. Compare your graph with Figure 3(b). How has the composition of the Canadian workforce changed during the 20th century? Explain what has caused these changes.

Turn to page 260 to review making circle graphs.

Industry	1997 (thousands)
Agriculture	453
Fishing, forestry, mining, trapping	325
Manufacturing	2 297
Construction	858
Transportation, utilities and communication	943
Trade (sale and distribution of goods)	2 526
Banking, finance, real estate	817
Other services	5 619
Public administration (government)	828
Unclassified	543
Total	15 209

Figure 4
Numbers of Canadians Working in Various Industries

Case Study   *How Canada Compares*

Between 1911 and 1997, many changes took place in the Canadian economy. These included much growth in the secondary and tertiary sectors. In 1911, fairly equal numbers of people were employed in each of the three industrial sectors (see Figure 3(b) on page 118). Over the next decades, the percentage of people employed in primary industries declined dramatically. At the same time, the percentage of people employed in tertiary industries rose sharply. More and more products came to be manufactured in Canada, replacing imported goods. Manufacturing has remained an important industrial sector.

Poland in 1911, like Canada, had roads and railways, valuable mineral, forest and fishing resources, and large areas of good agricultural land. But its economy did not develop as Canada's did. Its present-day employment structure, shown in Figure 5, is not much different from Canada's in 1911. Why does this nation, with so much potential, lag about 80 years behind Canada in its economic development?

The answer lies in Poland's history. Since the 17th century, Poland has been repeatedly invaded and almost continuously controlled by outside forces. From 1945 to 1989 it was controlled by a *communist* government. Under communism, many industries were very inefficient. Housing and food shortages were common. The government ("state") owned all of the country's property and companies, and there were no opportunities for entrepreneurs. The country's capital resources were limited, so

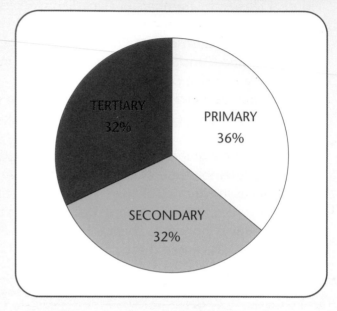

Figure 5
Employment in Poland, 1996

machinery and equipment wore down or became out-of-date.

If you visit Poland today, more than a decade after the end of communism, you can see a nation struggling to improve its traditional farming methods and food supply system. Outdated industries stand side-by-side with newer ones largely built by foreign investors. Inefficient state-run enterprises have been allowed to go bankrupt. Old hospitals, schools, and other services are slowly being upgraded. The economy is growing at a tremendous rate of 6 per cent per year, compared to Canada's growth rate of 3 per cent. We can expect Poland to one day take its place as a major partner in the European economy.

Discover For Yourself

1. What is similar about Canada in 1911 and Poland in 1996?
2. Why does Poland's economy lag so far behind that of Canada today?
3. In small groups, discuss your predictions for the future development of the Polish economy. Do you think that it will take more than 80 years for the Polish economy to catch up with the Canadian economy? Give reasons for your answer. What obstacles might slow down the process?
4. Draw a circle graph showing your prediction for the distribution of workers in the three sectors of industry in Poland 20 years in the future. Give reasons for the size of each of the sectors that you draw.

Locations of Industries

What factors affect where industries are located? For primary or resource industries, the most important factor is the availability of the natural resource. Secondary industries in Canada are concentrated in southern Ontario and southern Quebec (see Figure 6 on page 122). Six advantages, or "location factors," explain why there is more manufacturing here than elsewhere:

▸ There are good *transportation* links by ship, train, truck, and plane.
▸ Many different *raw materials* used to make products are available.
▸ There is a large *labour force* with all kinds of different skills.
▸ *Markets* for products are large in southern Ontario and southern Quebec. The market in the eastern United States is nearby. Other markets are easily reached by ship.
▸ There is plenty of *energy* available to run machinery.
▸ *Capital*, or money to purchase machinery, land and so on, is available locally.

Figure 6 shows the locations of manufacturing activity across Canada.

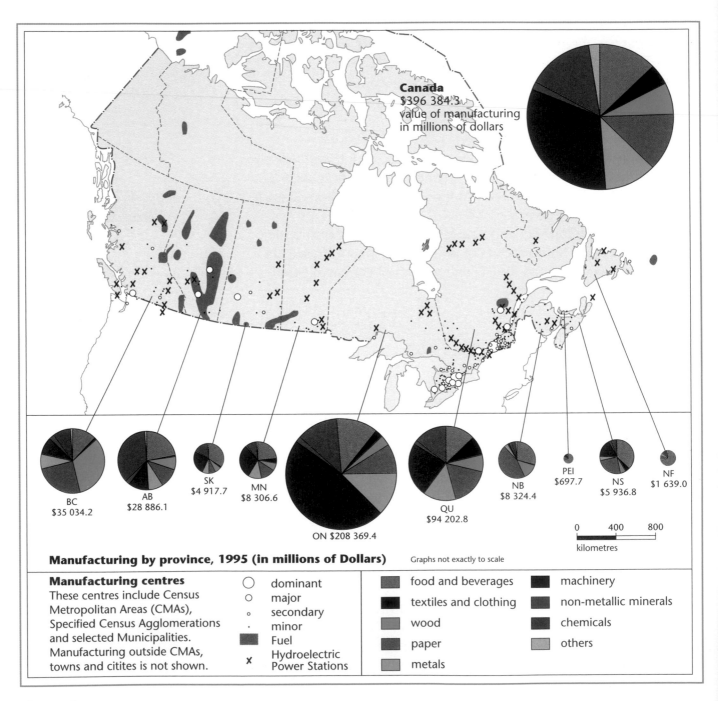

Canada
$396 384.3
value of manufacturing
in millions of dollars

BC
$35 034.2

AB
$28 886.1

SK
$4 917.7

MN
$8 306.6

ON $208 369.4

QU
$94 202.8

NB
$8 324.4

PEI
$697.7

NS
$5 936.8

NF
$1 639.0

0   400   800
kilometres

Manufacturing by province, 1995 (in millions of Dollars)   Graphs not exactly to scale

Manufacturing centres
These centres include Census
Metropolitan Areas (CMAs),
Specified Census Agglomerations
and selected Municipalities.
Manufacturing outside CMAs,
towns and citites is not shown.

○ dominant
○ major
∘ secondary
· minor
▪ Fuel
✕ Hydroelectric
    Power Stations

▪ food and beverages
▪ textiles and clothing
▪ wood
▪ paper
▪ metals

▪ machinery
▪ non-metallic minerals
▪ chemicals
▪ others

Figure 6
Manufacturing by Province, 1995

Tertiary industries provide services for people. As a general rule, therefore, the higher a place's population, the more service industries will be located there. For example, a city of 1 million people will have 10 times as many schools, hospitals, police officers, dentists, retail stores, restaurants, and snow ploughs as a town of 100 000. Geographers use the term *correlation* to describe the relationship between population and services.

Retailing is one of the most important parts of tertiary industry. About half of Canadians' total retail dollars are spent in food stores and car dealerships. The other half of this money is spent at gas stations, automotive centres, and drug, clothing, furniture and general stores. Figure 7 shows the number of people employed in tertiary industries in each province in 1997.

Turn to page 96 to review **correlation**.

Turn to page 15 to review the correlation between population and services.

Province	Population (thousands)	Number of Workers in Tertiary Industries (thousands)
Newfoundland	553	148
PEI	137	43
Nova Scotia	936	303
New Brunswick	753	215
Quebec	7 308	2 406
Ontario	11 254	3 959
Manitoba	1 139	401
Saskatchewan	1 023	335
Alberta	2 836	1 056
BC	3 964	1 421

Figure 7
Number of People Employed in Tertiary Industries and the Population of the Provinces in 1997

Discover *With Maps and Graphs*

1. Look at Figure 6 on page 122.
   a) Which manufacturing activity is important in every province?
   b) In which province do wood and paper manufacturing industries dominate the economy? Give reasons for this pattern using Figure 9 on page 125.
   c) Which type of manufacturing dominates in Ontario and Quebec?
   d) Rank the provinces from highest to lowest in value of manufacturing.
2. Look at Figure 7 on page 123.
   a) Make a scattergram to show the correlation between *population* (marked on the vertical axis, in thousands, from 0 to 12 000) and *number employed in service industries* (marked on the horizontal axis, in thousands, from 0 to 4000).
   b) What type of correlation does your scattergram show?

Figure 8
Why is the manufacturing activity shown here important in all Canadian regions?

Turn to page 263 to review making scattergrams.

Summary

In this chapter you have discovered how Canada's economy evolved from a resource-based system to one based on services and manufacturing. You have learned that industrialization has reached different levels around the world. You have also studied the factors that affect where industries are located in Canada.

Reviewing Your Discoveries

1. List four primary industries that have been important to Canada's economic evolution.
2. Give an example of a country with abundant natural resources that is less developed than Canada. Explain why the country has not evolved economically like Canada.

Using Your Discoveries

1. Imagine that you want to start up a factory to manufacture "Canadian Springtime Candy." You will need the following: raw materials (maple sugar, butter, and lemon juice); capital (machinery, cooking equipment, a factory building, land, and money); electricity (to run the machinery); and labour (three workers). You want to reach the largest possible market. Form small groups to work on the development of this factory.

   a) Research how maple sugar is made and where its suppliers are located.

   b) Research each of the four communities in Figure 9 for their suitability as locations for your factory. List the advantages and disadvantages of each community for each of the location factors in the top row of Figure 9. To help you make your list, refer to Figure 6.

   c) Choose the best location for your factory.

   d) What other questions would you want answered in order to make the best location choice?

Turn to page 122 and 29 to help you make your list.

Possible Factory Location	Transportation	Raw Materials	Labour Force	Market	Energy	Capital
Edmonton, Alberta						
Churchill, Manitoba						
Toronto, Ontario						
Mirabel, Quebec						

Figure 9
Fill in this chart to help you decide on the best location for your factory.

Chapter 3

Formulas for Success

Key terms

In this chapter we focus on the present-day success of the Canadian economy. The information and activities will help you

▸ show an understanding of how economic resources influence a region's economic success

▸ show an understanding of the manufacturing system

▸ analyze data about the economy of Canada and other G-7 member nations.

The Importance of Manufacturing

In the last chapter, we learned that Canada's economy has grown to include activity in three sectors. We saw that Canada's highly developed manufacturing and service sectors are key factors in the country's prosperity.

All of us have reasons to be thankful for Canada's manufacturing industries. Perhaps someone in your family is among the 15 per cent of Canada's workers employed in manufacturing (as of 1998). Perhaps there are manufacturing industries in your community. Such industries not only provide employment, but also pay high taxes. These taxes help pay for roads, parks, garbage collection, snow clearing and other services that keep your community safe and pleasant. When a city loses manufacturing industries, the only way to maintain these services is to raise the taxes that homeowners pay.

Manufacturing also benefits the country as a whole. The more Canada manufactures, the less it needs to import from other countries. When manufacturing is *mechanized*, or machine-run, a large amount of goods can be produced at low cost. The many manufactured goods available in Canada today help to raise the general standard of living in the country.

Figure 1
Manufacturing includes (a) cottage industries, where craftspeople work out of their houses producing small items, and (b) large machine-run factories, where people often work on an assembly line. How do Canadians benefit from these manufacturing industries?

GEO-TOOL

Factors of Production

Just as the physical world has *natural resources* (e.g., air, forests, minerals), countries have *economic resources*. These economic resources are also called "factors of production." The four factors of production are: *land, labour, capital* and *skilled entrepreneurs*. Economists use these factors to explain why some regions are more economically successful than others. Figure 2 gives some examples of how these four factors play a role in Canada's economic success. As you study the figure, think about how these factors help manufacturing industries.

Land

The Natural Resources Used to Produce Goods and Services

Examples:
▸ metals (from the Canadian Shield and mountains of the west and east)
  – used to make machines, cars, trucks, buildings, electrical wiring
▸ oil, natural gas, coal (from lowlands and plains)
  – used to produce energy for industry and transportation
▸ other minerals
  – used to make chemically-based products (e.g., fertilizers, plastics)
▸ gravel, sand
  – used to build roads and other structures

Figure 2
Vast quantities of gravel are needed to build and maintain our streets and highways.

Labour

The Time and Effort Used to Produce Goods and Services

Examples:
▸ finding and extracting natural resources (for primary industries)
▸ researching how to make better products more efficiently (for secondary industries, for health care in the tertiary sector)
▸ engineering and constructing public works (dams, bridges, highways)

A country's labour force needs to be well educated to function in the full range of jobs the economy provides.

Capital

The Equipment, Tools and Other Manufactured Goods Used to Produce Goods and Services

Examples:

- Buildings/Factories
- Machinery and equipment ⎫ "Stock"
- Supplies ⎭

One way of raising money for a business is to sell *stocks* or *shares*. The value of a company's stock (see above list) is divided into equal shares. People pay money to own one or more shares and become part owners in the company. They receive an appropriate share in the profits of the company and can sell their stock certificates at any time. At the Toronto Stock Exchange (above), trading in stocks takes place.

Entrepreneurial Ability

People's Ability to Organize the Other Three Factors of Production

Examples:

- Edward (Ted) Rogers, developer and owner of Rogers Communications Inc., a company involved in cable television, radio broadcasting, magazine publishing, and cellular phones
- Kwok Yuen Ho (winner of the Ontario 1998 Entrepreneur of the Year Award), who started ATI Technologies, a computer component company and the third largest high-tech firm in Canada
- Viola MacMillan, the "Queen Bee of Canadian Mining" and president of the Prospectors & Developers Association for 21 years

The Canadian Government encourages entrepreneurs to immigrate to this country. Upon arriving, they have two years to prove they are running a business that employs at least one Canadian who is not a member of their family.

Discover For Yourself

foreign investment—one country allowing people from other countries to invest, or put money into, its industries.

LANGUAGE LINK

1. Form small groups and discuss these questions:
   a) Choose any three school subjects. How could each of them benefit your working lives and, as a result, the economy?
   b) How could the education system be changed to better prepare young people for the world of work? Explain your ideas fully.
2. One way that Canada raises the capital needed to develop its businesses is to allow **foreign investment**. Canada's foreign investment laws were changed in 1984, with the result that direct foreign investment more than doubled over the next decade.
   a) Look at Figure 3. Why do you think our closest foreign investment ties are with the United States and the United Kingdom?
   b) Describe three advantages and three disadvantages of foreign investment in Canada. Use the words or phrases below in your answer.

head office control	*taxes*	*investor profit*
branch closures	*jobs*	*new technology*
3. In small groups, discuss whether or not the government should encourage large amounts of foreign investment in Canada. Give reasons for your opinions.
4. Complete the following statements to see if you would make a good entrepreneur. Check your answers on page 138.
   a) I mainly look to .... to solve a problem.
      i) friends   ii) experts   iii) myself
   b) I am motivated by ....
      i) achieving goals   ii) gaining attention   iii) controlling others
   c) Any business I ran would succeed or fail because of ....
      i) luck   ii) support from others   iii) my abilities
   d) My preferred job would be ....
      i) challenging, with some risks   ii) high-paying, with high risks
      iii) easy, with few risks

e) My preferred job would allow me to ....
   i) *be competent and efficient*   ii) *choose how to use my time*
   iii) *create and do new things*
f) Profits from a business I ran would be important because they would ....
   i) *be reinvested in the business*   ii) *show success*
   iii) *make me richer*

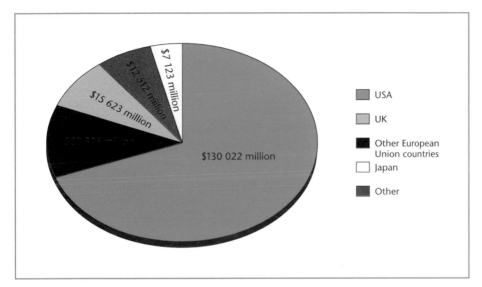

Figure 3
Sources of Direct Investment in Canada, 1997

Manufacturing Success

Every kind of manufacturing is made up of four elements. These are *inputs*, *processes*, *outputs*, and *feedback*. Even a simple manufacturing system, such as a roadside lemonade stand, includes these elements. In the case of the roadside stand, the **inputs** are *lemonade powder (or lemons)*, *sugar*, *water*, and *paper cups*. The process involves mixing the inputs together in the correct ratios. The **outputs** are the lemonade and used paper cups. The lemonade maker gets feedback by listening to customers—did they like the lemonade? Could it be made better? Is the price acceptable? Will they come back for more?

inputs—the factors of production put into a manufacturing system. Inputs get worked on by processes.

outputs—the products leaving a manufacturing system that result from processes.

Case Study   *Manufacturing Steel*

The Dofasco plant in Hamilton is an example of a successful operation involving all four elements of manufacturing. For making steel, the first set of elements—the inputs—include *coal*, *iron ore*, and *limestone* as main ingredients, as well as air, water, and electrical energy. The main ingredients come from North American suppliers. Water is available from nearby Hamilton Harbour. Electricity is available from the Ontario Electrical Grid, whose lines pass right by Hamilton as they carry energy from generators at Niagara Falls.

The process for making steel consists of several stages.

1. Coal is changed into "coke" by being heated in a coke oven.
2. Coke and iron are loaded into a blast furnace. They are heated and the liquid iron melts out from the ore. Limestone is added to the mixture in the blast furnace. It joins with the impurities in the iron ore and separates them from the molten iron, making the molten iron pure. (The combination of limestone and iron ore impurities is called "slag," which floats to the top and is skimmed off.)
3. The molten iron is mixed with scrap in an oxygen furnace to form liquid steel.
4. The steel is poured into a ladle where special chemicals may be added.
5. The steel then goes to the slab caster where it cools, solidifies, and is cut into 10 m lengths.
6. The slabs are then reheated and rolled.

Dofasco's Improvements

Between 1980 and 1995, Dofasco spent over $3 billion on technological improvements. Many

Figure 4
The daily output of Dofasco's process is 11 000 tonnes of steel, consisting of over 1000 types.

of the innovations have cut the cost of making steel and helped reduce waste and pollution. For example, a $200-million electric arc steelmaking furnace uses 70 to 100 per cent scrap steel. A continuous slab caster eliminates the time and space needed to make steel *ingots* (huge blocks of hardened steel). This technology has increased the speed and reduced the cost of steel production.

Other changes resulted when Dofasco employees searched for and found solutions to meet their customers' special needs. A new mill produces steel tubes for the truck and car parts industry. At the parts maker, these special tubes are slid inside hollow forms and then filled with high pressure water. The water forces the steel to expand to the shape required. This process is called "hydroforming."

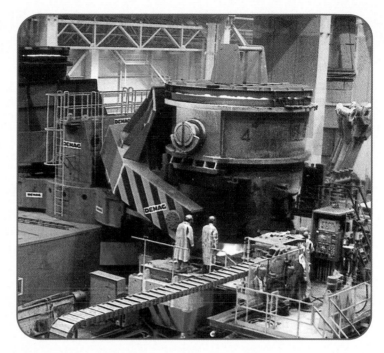

Figure 5
Dofasco's Continuous Slab Caster

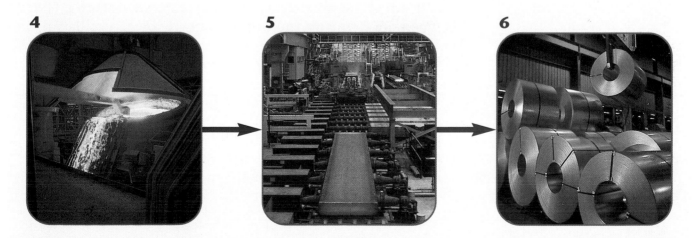

 *For Yourself*

1. In small groups, summarize steel manufacturing by answering these questions:
   a) List six inputs for steelmaking.
   b) Divide the steelmaking process into four main stages. Name the equipment used in each stage.
   c) List 20 products made from the outputs of the manufacturing process.
   d) How do you think Dofasco gets feedback from its customers? What changes have taken place as a result of this feedback?
2. List the similarities and differences between running a lemonade stand and running a steelmaking company. Use each of the four factors of production in your answer.

The Proof of Success

Canada's strength in manufacturing and its consistently growing economy have allowed it to become a member of an organization called the "Group of Seven" ("G-7" for short). This organization is made up of the world's most **industrialized nations**. Other members are Germany, the United States, France, Italy, Japan, and the United Kingdom. Russia was invited to join the group in 1994.

The organization was originally formed to make decisions about international economic, financial, and trade policies. Today, however, it examines other important topics such as arms, energy, foreign debt, and technology.

Canada's membership in this influential group speaks well for its economic success. The only grave economic setback the country experienced in the 20th century was from 1930 to 1933,

industrialized nations— countries that use high levels of technology in all sectors of the economy.

WEB LINK **To learn more about Dofasco steel manufacturing, look up http://www.dofasco.ca**

when Canada's Gross National Product (GNP) decreased each year. During this period, part of the worldwide "Great Depression," millions of people were out of work, and the population generally became much poorer. But Canada's total production did rise in the decade as a whole, as shown in Figure 6.

Year	GNP in Millions of 1986 Dollars	% Growth in Previous Decade
1930	57 086	—
1940	76 864	34.6
1950	134 984	
1960	225 551	
1970	385 583	
1980	628 293	
1990	757 029	
1998	886 170	17.3 (pro-rated from 8 years for the decade)

Figure 6
Canada's Gross National Product (GNP) 1930–1997 (Based on the Value of the Dollar in 1986)

Figure 7
The Asian economic downturn affected Canadian exports to Indonesia, shown here. Exports fell dramatically in 1998, during a time when much of Indonesia's population could not afford food.

Canada's economy faced a second challenge in the 1990s when many Asian nations, including Japan, experienced a serious economic downturn. These nations are important markets for Canadian exports of raw materials such as coal and timber. The downturn meant that the Asian nations could not afford to buy as many of our products as before, which affected our economy. In Canada there were fewer jobs, and less money was available, so the economy slowed down, although it still continued to grow.

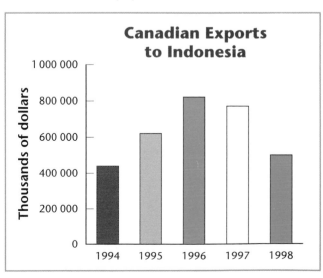

Canadian Exports to Indonesia

Discover With Graphs

1. Look at Figure 9.
    a) Write two statements that explain how Canada's population compares with that of other G-7 nations.
    b) Plot a scattergram relating GNP to population. Put "GNP ($US billions)" on the vertical axis and "Population (millions)" on the horizontal axis. Write the name of each country beside its dot. Plot the dot for Russia, but do not include it when judging where to draw your straight line. Finish your graph by labelling your axes and including a title.
    c) With the exception of Russia, what is the relationship between population and GNP? How close is the relationship?
    d) Why would you expect this relationship to exist? (Hint: think about the definition of GNP.)
    e) Russia's economy is faced with many problems, including the same economic changes that have affected Poland since 1989. How would these changes affect the productivity of the country? How would they affect the income of each person? Explain your answers.

Turn to page 120 to review
Poland's economic
conditions.

Country	1999 Population (millions)	1997 GNP ($US billions)	1997 GNP per person ($US)
Canada	30.6	595	19 640
France	59.1	1 489	26 300
Germany	82.0	2 322	28 280
Italy	57.7	1 160	20 170
Japan	126.7	4 812	38 160
Russia	146.5	395	2 680
UK	59.4	1 229	20 870
USA	272.5	7 788	29 080

Figure 9
Populations and GNPs of
G-7 Member Nations

f) Give two reasons why you think the G-7 nations invited Russia to join their organization, even though its GNP is very low for the size of its population. (You may need to ask some adults about Russia's background before 1990.)

2. Look back at Figure 6 on page 135.

a) Calculate the percentage by which the GNP grew in each decade shown in Figure 6. The first one (growth from 1930 to 1940) and the last one (growth from 1990 to 1998) have been done for you. To calculate the growth from 1940 to 1950, follow the formula below:

   (*GNP 1950 — GNP 1940*) x 100 ÷ *GNP 1940*
   (134 984   — 76 864)   x 100 ÷ 76 864

b) During which four decades was growth in the Canadian economy especially strong?

c) When the economy grows quickly, many jobs are created and people's incomes tend to rise. How would slow economic growth have affected people in the 1930s and 1990s?

Figure 8
Russia attended its first G-7 summit as a full partner in 1997 when the meeting was held in Denver, U.S.A.

Summary

In this chapter you have learned about the importance of manufacturing and factors of production to the Canadian economy. You have discovered how manufacturing works and how it has progressed in the case of Dofasco steel. You have also seen the proof of Canada's economic success in its growing economy and international influence.

Reviewing Your Discoveries

1. Why is manufacturing an important part of the Canadian economy?
2. What are the four stages involved in all manufacturing operations?
3. Describe the four main factors necessary for economic success.

Measure Your Entrepreneurial Ability

a) The best choice is (ii). Even though entrepreneurs are independent, they know they need the help of others.

b) The best choice is (i). Many entrepreneurs are most motivated by their need for personal achievement.

c) The best choice is (iii). Many entrepreneurs believe strongly in themselves, their choices, and their abilities.

d) The best choice is (i). Success is important to entrepreneurs, so they don't want too many risks. At the same time, they don't want success to come too easily.

e) The best choice is (iii). Research has shown that (i) is the choice of business managers, (ii) is the choice of college professors, while entrepreneurs are most interested in (iii).

f) Choices (i) and (ii) are better than (iii). Entrepreneurs are committed to building successful businesses. Making profits is a means to success and growth, not an end in itself.

Turn to page 62 to review location factors.

Using Your Discoveries

1. Form small groups and research a manufacturing industry in Canada today. Present your findings to other groups. Include location maps, flow charts, graphs, and other visual aids in your presentation. Your findings should include the following:

   a) the name of the company, its location(s) and major product(s)

   b) an explanation of the location factors that influenced where the firm chose to locate

   c) the operation's inputs, process and product, with an explanation of how the firm gets feedback

   d) a short history of the firm

   e) the number of employees and kinds of jobs people have in the firm

   f) the advantages and disadvantages that would influence whether you would want to work in the firm.

2. The North American space program has resulted in many technological **spinoffs**. These spinoffs have had a tremendous impact on the economic systems of industrialized nations. Some of them are shown in Figure 10.

spinoff—a product or an idea resulting from an activity that was designed for a different purpose.

   a) Choose any three spinoffs. For each, explain why it was developed for the space program.

   b) Reflect on these spinoffs and then write a few paragraphs describing how the space program has affected
     – our daily lives
     – the way we run our businesses
     – the methods used to manufacture products
     – our understanding and management of the environment.

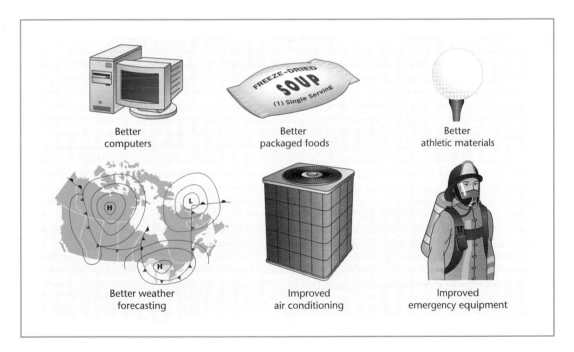

Figure 10
Space Program Spinoffs

To find out more about benefits from the space program, look up http://www.thespaceplace.com/nasa/spinoffs.html

Chapter 4

Comparing Economies

Key terms

subsistence economy

traditional economy

command economy

subsistence economy—an
economic system in which people's
labour only produces enough
food, clothing, and shelter for the
workers' own needs.

In this chapter we focus on the different kinds of economies found around the world. The information and activities will help you

▶ show an awareness of the characteristics of different economic systems

▶ recognize that mixed economies are widespread

▶ analyze the economies of three Ontario communities.

The Earliest Economic System

Economic systems have existed for as long as people have put effort into meeting their needs. In some parts of the world, people work in economic systems that have not changed for hundreds of years. In most of these cases, getting food, fuel, and other items needed for survival is the focus of people's economic activity.

Subsistence economies are widespread in Africa, South America and central Asia, where people plant small amounts of crops or herd animals in dry climates with poor soils. They are also found in areas of tropical rain forest, where Aboriginal peoples subsist on the forest's wood, fruits, roots, medicines, and meat. Unfortunately, many of these economies are being threatened by other human activity in the environment.

Much of the land of the Penan people in Sarawak (an island in Southeast Asia) has been spoiled by loggers. Pollution from gold mining is destroying the subsistence economy of the Yanomami people, living along the Venezuela–Brazil border.

Traditional Economies

Activity in a **traditional economy** is a little more productive than subsistence activity. Some rice farmers in Southeast Asia grow enough rice to feed their families and have some left over to trade for other goods. Inshore fishing is also common in traditional economies. Traditional fishers are threatened by the more efficient methods of larger fishing vessels. These larger operations produce fish at lower costs, so fishers in small boats may find it difficult to compete. Another problem is that overfishing in areas such as the waters off Newfoundland has depleted fish stocks to a great extent. The photos below show examples of traditional and subsistence economies.

traditional economy—an economic system in which people's methods of working have changed little from one generation to the next. Workers in a traditional economy try to produce a little more than what is needed for subsistence.

Figure 1
Subsistence agriculture in Morocco (a) and slash-and-burn farming in Sarawak (b) are not as productive as the traditional economic activities seen in (c) and (d).

1. Discuss in small groups whether each of the following activities is likely to be part of a subsistence economy, a traditional economy or neither. Give reasons for your choices.
   a) hunting and fishing
   b) making electronic equipment
   c) quilt making
   d) slash-and-burn farming (clearing a small area of forest and planting crops)
   e) vegetable farming
   f) the Summer Olympic Games
2. Describe a situation in which a traditional economy is also a subsistence economy. Be specific in your description.
3. In some developing countries, a large part of the population makes its living in a subsistence economy. Explain why these countries report an extremely low GNP per person.

Organizing a National Economy

Many regions of the world have some subsistence or traditional economic activity. But even in the developed world, there are still areas where traditional and/or subsistence economies remain. For example, some remnants of earlier traditional economies in Canada still exist in remoter areas where hunting, trapping and fishing still form an important part of people's lives. However, countries as a whole tend to require more complex economic systems. These systems involve the *government*, *business/industry*, and *consumers*.

In today's world, there are three general forms these systems may take. At one extreme the government takes total control of business/industry, preventing any influence by consumers. At the other extreme, the government gives business/industry and consumers total freedom to organize themselves, so that they interact and influence each other. A third type of system combines some features of both extremes.

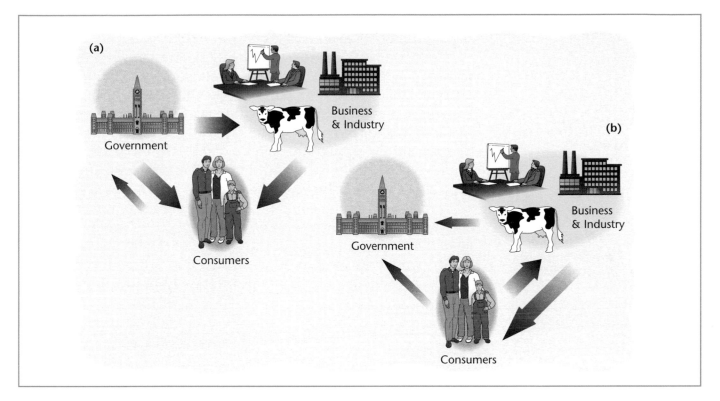

Figure 2
In which economic system does the government make all of the key decisions? In which economic system do business and industry respond to consumer demand for goods and services? Which economic system do you think gives the best opportunities to entrepreneurs?

Command Economies

Command economies (Figure 2(a)) are found today in communist nations. This system originated in Russia in 1917 after the Russian Revolution, but many of its ideas were developed by a German living in England in the 19th century, Karl Marx. Marx believed that in an ideal world, all people would work as well as they could for the sake of the country, and in return they could take from the country whatever they needed. This would result in a decent standard of living for everyone, with nobody becoming very rich or very poor. Communism and the command system spread to other countries, such as China, Poland and Cuba, between 1945 and 1975. In 1989, many of the communist countries in Europe

command economy—an economic system in which the government owns and controls all parts of the economy.

abandoned both communism and the command economy, but the system still exists in other parts of the world.

In this system, all property, including farmland and industries, is owned by the state. The state pays all workers, and almost everyone has a job and steady income. On the downside, most workers in a command economy find that there is no reward for working hard or trying to improve the system. There is no competition among similar industries so industrial methods do not improve quickly. Construction is often shoddy, and many buildings, roads, and machines deteriorate very quickly. Agriculture is not as productive as it could be, leading to food shortages.

The Chinese economy is one of today's surviving command economies. The country is in the process of modifying the system by allowing more local decision-making and private ownership. This has resulted in a tremendous increase in production.

Figure 3
In the 20th century, 20 countries, or "republics," in Europe and the former USSR had a command economy for anywhere from 40 to over 70 years.

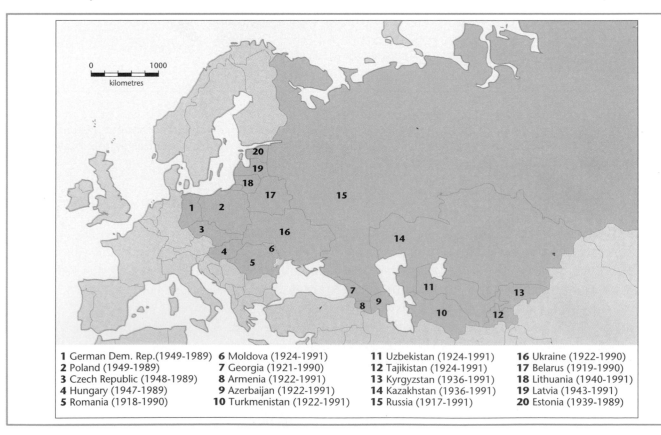

1 German Dem. Rep.(1949-1989) 6 Moldova (1924-1991) 11 Uzbekistan (1924-1991) 16 Ukraine (1922-1990)
2 Poland (1949-1989) 7 Georgia (1921-1990) 12 Tajikistan (1924-1991) 17 Belarus (1919-1990)
3 Czech Republic (1948-1989) 8 Armenia (1922-1991) 13 Kyrgyzstan (1936-1991) 18 Lithuania (1940-1991)
4 Hungary (1947-1989) 9 Azerbaijan (1922-1991) 14 Kazakhstan (1936-1991) 19 Latvia (1943-1991)
5 Romania (1918-1990) 10 Turkmenistan (1922-1991) 15 Russia (1917-1991) 20 Estonia (1939-1989)

Market Economies

A true **market economy** (Figure 2(b) on page 143) allows anyone to create a business without help or interference from the government. A system where people are free to try any economic activity they want to try was first described by the Scottish philosopher Adam Smith in the 18th century.

Smith believed that this freedom could still produce an orderly economic system because an "invisible hand" leads even the most selfish businesspeople to work for the good of society. Smith said that competition between businesses is one way society benefits from the invisible hand. For example, even when businesspeople are working only for their own benefit, competition keeps them from overcharging for their products, underpaying their employees, or taking advantage of people in other ways.

In a market economy, those businesses producing goods or services that customers want will flourish. Others will die because of lack of demand. Thus, the market determines how the economy will run and what it will produce. In such an environment, an entrepreneur with a good idea and lots of energy has an excellent chance of succeeding.

No nation in the world has an economic system *totally* controlled by the market. The United States, however, comes close. Fully two-thirds of the United States' GNP is generated by privately owned businesses. Only one-sixth of the GNP is the result of government activities such as transportation, the military, and education.

The government's involvement in the economy is limited to

(1) managing money policies and

(2) stepping in to help faltering companies if their failure would have a serious effect on the economy.

An example of the second kind of involvement occurred in 1980, when the American government loaned money to the Chrysler Corporation. Generally, however, this kind of government help is rare in the United States.

market economy—an economic system in which private individuals set up, own, and direct businesses that produce goods and services that consumers want. This system is also called "free enterprise" or "capitalism."

For Yourself

1. Look at Figure 4.
   a) Make a copy of the figure. Fill in the third column using the information in Figure 3 on page 144. What pattern do you see?
   b) In Russia and Romania, there has been little private ownership of businesses and farms for over 70 years. Why would this be a problem as these countries try to make a transition to a market economy?

Country	GNP/Person, 1997 (US$)	Years Spent in a Command Economy
Czech Republic	5240	
Poland	3590	
Romania	1410	
Russia	2680	

Figure 4
GNPs of Selected European Nations

2. Give three reasons why a market economy is much more attractive to an entrepreneur than a command economy.
3. The *black market* is common in command economies. In market systems, an *underground economy* flourishes because people have to pay high taxes to buy certain goods and services.
   a) Find out what the terms "black market" and "underground economy" mean and write out their definitions.
   b) Discuss in small groups why each of these illegal activities is common in its particular economic system.

Mixed Economies

Canada is one of many countries around the world with a **mixed economy**. Our governments have quite a lot of influence in some parts of the economy, but very little in others.

Some aspects of agriculture in Canada are under considerable government control. For example, the government controls the amount of milk and eggs that are produced by farmers. This controls the supply of these products and keeps prices stable. In addition, certain crops are sold at agreed prices to prevent farmers from competing with each other and putting each other out of business. The government sometimes comes to the aid of farmers when poor market prices threaten their livelihood. The government may also step in when bad weather damages or destroys farm crops.

Other areas the Canadian government invests in include education, job training, national defence, policing, road building, health care, employment insurance, welfare, and pensions. The government also provides help for those who are trying to start a business. Local governments (the level of government in charge

mixed economy—an economic system that combines private ownership with government control.

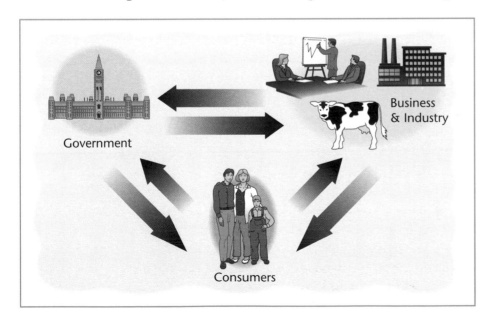

Figure 6
A Mixed Economy

of matters in the community or city) may encourage industries to come to their city or town by cutting taxes.

In mixed systems, government involvement in the economy varies depending on which party is in power. Conservative governments tend to encourage business development by cutting business taxes. They may also lessen controls on environmental protection in order to cut the production costs of companies. This is because they believe that the more profit companies make, the more the economy grows and the more people are employed. Conservative governments usually spend less money on state-run programs, such as schools, hospitals, and social services. As a result, richer people have better opportunities than the poor, who cannot afford to buy services not provided by the state.

At the other extreme, socialist governments tend to tax businesses more heavily. This gives the government more money for state-run services. But industrial growth is hampered by these heavier taxes, so the economy may suffer. Unemployment often rises. Excessive spending by socialist governments may force them to borrow money and go into debt. The interest on debts becomes another obstacle to economic growth.

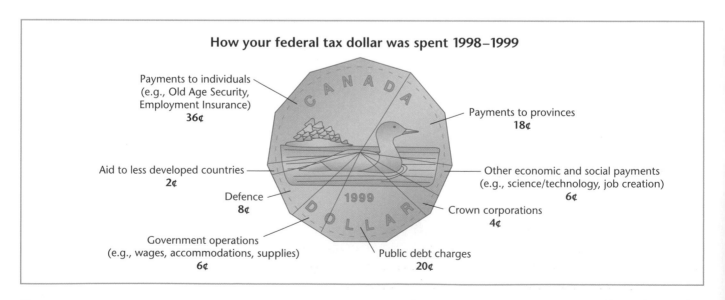

How your federal tax dollar was spent 1998–1999

Payments to individuals
(e.g., Old Age Security,
Employment Insurance)
36¢

Payments to provinces
18¢

Aid to less developed countries
2¢

Other economic and social payments
(e.g., science/technology, job creation)
6¢

Defence
8¢

Crown corporations
4¢

Government operations
(e.g., wages, accommodations, supplies)
6¢

Public debt charges
20¢

Figure 6
How do you think the government decides how it will split up the tax money it receives each year?

Discover ☼ For Yourself

1. Why do governments in Canada
   a) control the production of milk and eggs?
   b) help businesses get started?
2. Discuss the following two questions in small groups. Give reasons for the position you take.
   a) Do you favour the conservative or the socialist approach to organizing the economy?
   b) The federal government (the level of government in charge of the country) and provincial governments (the level of government in charge of a province) change frequently, and economic policy changes with them. Do you think this is a good or a bad thing?

Mixed Economies in Towns and Cities

Most communities need the support of local governments to keep their economies strong. Business and industry also need to be involved in employing as many community residents as possible. A survey of Ontario communities shows great variety in the types of businesses found in the community and the types of local government involvement. In most cases, both business and government want the economy to be as *diversified*, or broadly based, as possible.

Elk Lake is an example of a very small and not very diverse economy. The logging industry and planing mill are the economic cornerstones, employing about 200 workers. The total population is between 450 and 500. Only in such a small community could the economy depend so much on just one natural resource.

Kirkland Lake is a larger community of about 10 000. Although gold mining is very important to the economy, the community's local government has been trying very hard to attract more manufacturing to the area. Businesses are offered grants and loans from all levels of government, as well as very reasonable land prices, to settle in Kirkland Lake. A controversial project being considered is using an abandoned mine to store solid waste that would be shipped by rail from Metro Toronto.

In contrast to these northern Ontario communities, Guelph has a much larger population of 90 000. It is located within commuting distance of many large cities and offers employment in a great range of occupations. One of its large firms, the McNeil Consumer Products Company (makers of Tylenol), employs about 375 people, many in high technology research and development. The city hosts a university as well as a great variety of manufacturing companies. Figure 7 summarizes the economic mix in these diverse communities.

| | % Workers in | | |
Community	Primary Industries	Manufacturing and Construction	Service Industries
Elk Lake	26	28	46
Kirkland Lake	15	10	75
Guelph	2	31	67

Figure 7
Labour Force in Different Sectors of Industry in Selected Ontario Communities, 1996

Figure 8
Kirkland Lake, Ontario

Figure 9
Guelph, Ontario

For Yourself

1. Get or make an outline map of Ontario. Using an atlas or road map of Ontario, mark with red dots the locations of the three communities in Figure 7 and label them in black printing.

   a) For each community, draw a circle graph to show the percentage of workers employed in each sector of industry. Make your circle graphs different sizes to represent the different sizes of the workforce in each community. Follow these guidelines for the *radius* of each circle graph: Elk Lake— 9 mm; Kirkland Lake—39 mm; Guelph—129 mm. Use the same colours in all three graphs.

   b) Carefully cut out each circle. Arrange the circles around the outline map of Ontario, drawing straight lines that do not cross from each graph to its location on the map. Glue the map and graphs to a larger piece of paper. Finish your map with a north sign, neat legend, and suitable title.

2. Look at your circle graphs and the locations of the communities on your map.

   a) What differences do you notice between the two northern Ontario communities and Guelph in southern Ontario?

   b) What factors do you think account for these differences?

3. Imagine that the primary industries in each community shut down.

   a) Rank the communities based on the impact of the shutdown (from most to least impact).

   b) Give two reasons why the loss of 100 jobs in a northern community is much more significant than the loss of 100 jobs in a southern community.

Summary

In this chapter you have learned about five different economic systems. You have discovered the advantages and disadvantages of each. You have also seen some ways in which the economies of Ontario communities differ.

Reviewing Your Discoveries

1. What kind of economic system is described in each of the following statements?
   a) The government makes all of the decisions.
   b) The people have just enough to survive.
   c) Government makes some decisions, but private businesses make many others.
   d) The economic activity is the same as has existed in the area for 200 years.
   e) Businesses respond only to consumer demand.

Using Your Discoveries

1. In small groups, research the economy of your own community. Find out the kinds of industries present and the types of employment opportunities available locally. Include information on the balance among the three different sectors of industry. Present your findings as a short report that includes
   a) one paragraph of writing
   b) one graph
   c) two lists or tables

2. Decide as a group on three proposals to improve the local economy.
   a) Do you want government or businesses to make the improvements?
   b) Can you think of any problems that may develop as a result of your proposals?

Chapter 5

The Value of Trade

In this chapter we focus on trade between Canada and other nations. The information and activities will help you

▸ describe the advantages and disadvantages of economic associations
▸ describe the economic relationship between Canada and the global community
▸ identify top trading countries of the world.

Imports and Exports

A great deal of the products you use every day are likely to be **imports**. For example, the clothing you put on in the morning may have been made outside Canada. So might the radio you listen to as you dress. The orange juice you drink for breakfast came from a southern climate outside Canada, since Canada is too cold to grow oranges.

Canada is also too cold to produce the ingredients of the chocolate bars and cola drinks you might have for a snack. Your parents' car may have been imported, as well as the computer you might use at school or home.

One reason we buy products from foreign nations is because we do not make those products in Canada. Another

Key terms

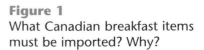

imports
exports
North American Free Trade Agreement

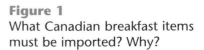

imports—things brought into a country.

Figure 1
What Canadian breakfast items must be imported? Why?

reason is because they are cheaper than Canadian products. Or we might like them better than Canadian-made goods.

We also import services, which make up about 17 per cent of our total imports. These services include foreign-owned banks, insurance companies, consulting firms, entertainment, travel packages, hotels, and restaurants. We commonly pay for the labour of people in other nations. For example, data processing and paperwork for insurance claims are often done outside of Canada. The reason for this practice is that labour is cheaper in these other nations.

The cost of labour, or the amount of money paid to someone to do work, can vary enormously from country to country. Workers in developed nations earn the most on average, while those in developing nations earn the least. The graph below compares the amount a worker earning $100 in the United States would be paid to do the same work in eight other countries.

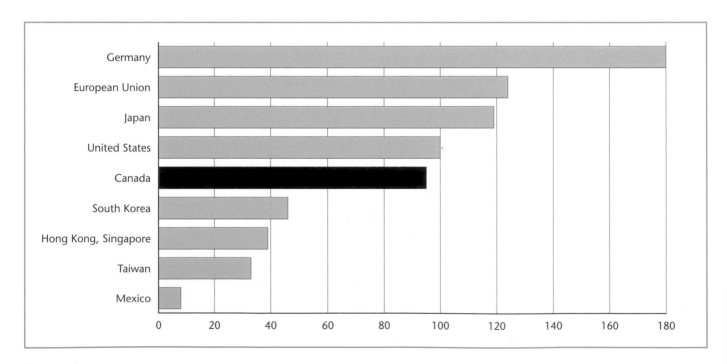

Figure 2
This graph shows that in 1996, a worker earning US$180 in Germany would only earn US$7 for doing the same work in Mexico. Why do you think the difference is so great?

In order to pay for imports from other countries, we must sell **exports**. These include agricultural products, natural resources, manufactured goods, and services. A large part of our exports consists of raw materials such as wheat, metals, wood, and fish. This is very unusual for a highly industrialized nation—most other industrialized nations import a great deal of raw materials and export very few. Services make up about 12 per cent of the value of our exports.

exports—things sent out of a country.

Canada's International Trade

The quality of our lives as Canadians is greatly affected by two factors: (1) our ability to produce the goods and services we need, and (2) our ability to buy the goods and services we cannot produce ourselves. In the past, when transportation systems were much less advanced, we produced most of what we needed for ourselves. We even put **tariffs** on many of the goods we imported from other nations to discourage Canadians from buying them when they could buy the same products made in Canada.

tariff—a tax put on an import. By making imports more expensive, tariffs protect a country's industries and jobs.

Today, however, efficient transportation and communication make international trade much easier and cheaper than before. Groups of countries have joined together in trading agreements that push the volume of trade to ever higher levels. These agreements reduce tariffs or, in some cases, get rid of them altogether.

The most influential trading agreement to which Canada belongs is the **North American Free Trade Agreement (NAFTA)**. It is no surprise, then, that the United States is our main trading partner. We also trade a great deal with the other highly industrialized areas of the world, namely Japan and the European Union.

North American Free Trade Agreement (NAFTA)—a trade agreement signed by Canada, the United States and Mexico.

By comparing the amounts of total exports and imports for 1998 in Figure 3 on page 156, you can see that the value of our exports is greater than that of our imports. This means that, overall, we had a *trade surplus*. Such a surplus helps the Canadian economy stay healthy.

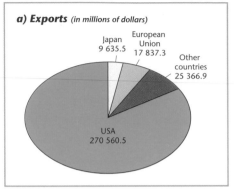
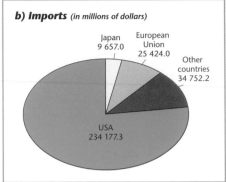

a) **Exports** *(in millions of dollars)*

Japan 9 635.5

European Union 17 837.3

Other countries 25 366.9

USA 270 560.5

b) **Imports** *(in millions of dollars)*

Japan 9 657.0

European Union 25 424.0

Other countries 34 752.2

USA 234 177.3

Figure 3
Value of Trade in Goods Between Canada and its Main Trading Partners, 1998

 **For Yourself**

1. Find 15 imported items that have their country of origin marked on them.
   a) In small groups, draw up a master list of all your items and their countries of origin.
   b) According to your list, which countries provide Canada with a large proportion of imported goods? Suggest reasons for this.
2. Why do you think Canada imports more services than it exports? (Hint: most of the cost of services comes from wages paid to workers.)
3. Figure 4 provides information about the kinds of goods that Canada exported and imported in 1998.
   a) Show the information in two circle graphs. Use the same colours or shading for the same items in each graph. Add titles and labels to finish your graphs.
   b) What percentage of our exports is made up of raw materials (the first four categories)? How does this compare to our imports of the same goods?
   c) Suggest two reasons why we would both import and export items in the same categories.
   d) Which raw material is exported much more than it is imported? Suggest two possible reasons for this.
   e) What is the difference between the values of our export and import of machinery and equipment? How might this be linked to our exports of raw materials?

MATH LINK

Commodity	Exports (%)	Imports (%)
Agriculture and fishing products	8	6
Energy products	7	3
Forestry products	11	1
Industrial goods and materials	18	20
Machinery and equipment	24	33
Automotive products	24	22
Other consumer goods	4	11
Other	4	4
Total	100	100

Figure 4
Main Goods Imported and Exported in Canada, 1998

Trading Agreements

Why do nations enter into trading agreements? What are their advantages and disadvantages?

Trading agreements make it easier for each member nation to sell its products to other member nations. This helps the economies of the exporting nations to grow.

There may also be disadvantages to trading agreements. But, compared to the advantages, they are less obvious and harder to prove. Take the case of Canada's most important trading agreement. On January 1, 1989, Canada and the United States became partners in a free trade agreement (trade with no tariffs). On January 1, 1994, Mexico joined Canada and the United States in what is now called NAFTA.

Problems have arisen because of differences in working conditions among the NAFTA countries. Canadian workers receive much higher wages than workers in Mexico (and workers in some parts of the United States). Canadian

Figure 5
The Countries of NAFTA

workers also have more rights and benefits than most workers in Mexico and some workers in the United States. Canadian laws force manufacturers in Canada to control their pollution levels, which increases Canadian production costs. Laws to control pollution in Mexico are rarely enforced.

As a result of these differences, Canadian products are more expensive than many products made in Mexico and the United States. Once tariffs are removed from the cheaper goods, these imports become very attractive to Canadian consumers. Consumers usually choose to buy more of the imports than they do of Canadian-made items. As a result, many Canadian factories have closed and workers have lost their jobs.

There are still many adjustments to be made before NAFTA operates fully. It is quite likely that other nations from Central and South America will join the United States, Canada, and Mexico. Only time will tell whether this agreement has or has not been beneficial for Canada.

The European Union (EU)

The European Union started with a 1951 agreement involving steel production between France, West Germany, Italy, the Netherlands, Belgium, and Luxembourg. Today, with the addition of other countries, the European Union consists of 15 member nations among which goods, services, people, and capital move freely. Figure 7 on page 161 shows the nations that make up this powerful, well-organized economy. Many European countries that are not members of the EU have trading agreements with it.

People living in EU countries use a common passport. Since January 1, 1999, some use the same currency, the Euro (€). The union has set up four different funds to raise the standard of living of those in poorer member nations.

Foreign and security policies are now jointly decided on in the union in addition to economic matters. For example, the European Parliament, which has representatives from all member nations, creates laws and policies governing all the nations. Thus, the European Union is both an influential economic organization and a large political entity.

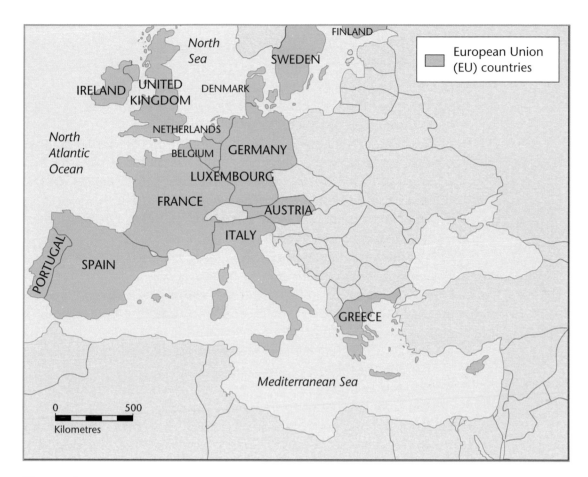

Figure 6
Countries of the European Union (EU)

Forming the union has not been easy. Many European nations have chosen not to belong. Some member nations have been slow to adopt the union's policies. Many people in member nations resent the loss of their national independence. Still, the streamlining of economic activity has benefitted the members of the union. The smaller members, especially, recognize that they now participate more actively in the global economy than they did before. The EU accounts for over 20 per cent of world trade, and it is likely that other nations will become members in the near future.

  **For Yourself**

1. Divide into groups to analyze and discuss these questions:
   a) What are three advantages of a country belonging to economic unions such as the EU and NAFTA?
   b) What are two disadvantages that result from economic unions?
   c) The European Union may become a model for other groups of countries linked by trading agreements. How would North America be affected if it followed the EU model?
   d) How do you think a closer association with the United States and Mexico could affect your lives?
   e) Would you be in favour of this closer association? Give reasons for your answer.

2. Figure 7 summarizes trading patterns within and between four areas of the world. Divide into groups to study the figure and answer these questions:
   a) Which of the four regions has the most internal trade? Which has the least?
   b) Suggest two reasons why there is so much internal trade in the area you identified in (a).
   c) Rank the trading regions (from largest to smallest) according to the value of the exports passing between them. Include the actual value of the exports for each. Your first item should be:
      – From Asia and Oceania to Western Hemisphere: US$333 billion
   d) Name two regions of the world that are most important in supplying goods and services to:
      – Western Hemisphere
      – Europe
      – Asia and Oceania
   e) Which region has very little trade with the rest of the world? What does this suggest about the region's level of industrialization and standard of living?

3. Figure 8 on page 162 shows global trade in microelectronics. Microelectronics includes computers, CD players, radios and televisions, as well as the small components used in these products.

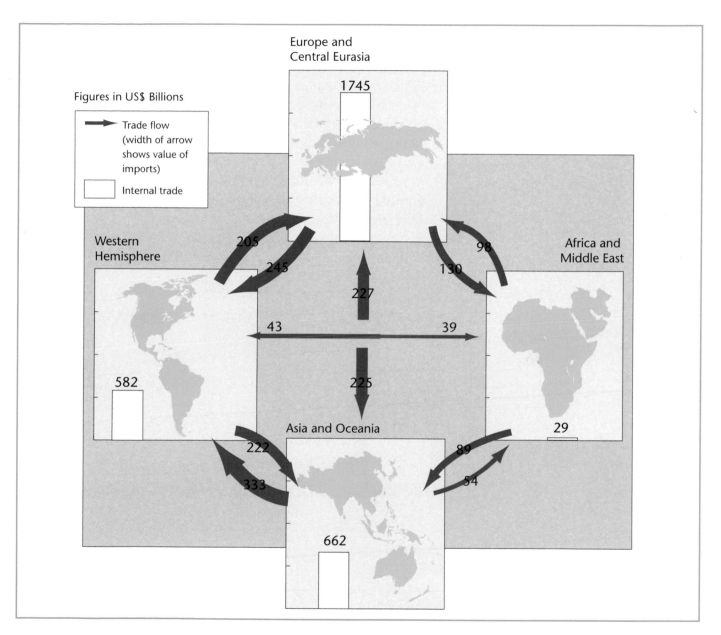

Europe and
Central Eurasia

1745

Figures in US$ Billions

Trade flow
(width of arrow
shows value of
imports)

Internal trade

Western
Hemisphere

Africa and
Middle East

205

245

227

98

130

43

39

582

225

29

Asia and Oceania

222

89

333

54

662

a) Make a copy of Figure 9 on page 162 and fill it in. To fill in the
middle column, add up the figures in all the import arrows
pointing to a region. To fill in the rightmost column, add up the
figures in all the export arrows pointing away from a region. For
example, Japan has two import arrows, labelled 3.67 and 5.80.

Figure 7
World Trade Within and Between
Four Regions, 1996

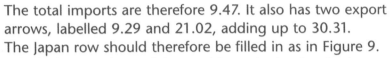

The total imports are therefore 9.47. It also has two export arrows, labelled 9.29 and 21.02, adding up to 30.31.
The Japan row should therefore be filled in as in Figure 9.

b) Which three regions of the world are the leaders in the export of microelectronic equipment?

c) Is Canada a net importer or net exporter of microelectronics? (A net importer imports more than it exports, while a net exporter exports more than it imports.)

d) In which regions of the world do you think most of the finished electronic equipment will be sold to consumers? Give reasons for your answer.

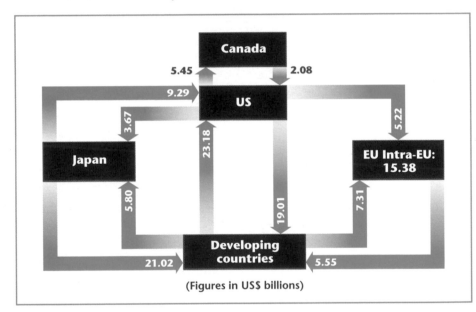

(Figures in US$ billions)

Figure 8
Global Trade Flows in Microelectronics, 1995

Region	Imports (US$ billions)	Exports (US$ billions)
Canada		
USA		
Japan	9.47	30.31
EU		
Developing countries		

Figure 9
Make a copy of this chart and fill it in to answer Question 3.

Case Study — The Asian Flu

In the mid-1990s, a series of events occurred that showed how complex global trade links can be. The nickname for these events was the "Asian flu." They revolved around a sudden decline in the value of currencies in many Southeast Asian nations.

To understand the situation that led to the crisis, we must go back to the year 1975 when the Vietnam War ended. At this time, Japan started to invest heavily in the manufacturing sectors of four Southeast Asian countries: Malaysia, Korea, Indonesia, and Thailand. Japan hoped that the manufactured exports from these nations would make the entire region a more important "player" in the global marketplace. The United States hoped that the development of the market economies of these nations would prevent the spread of communist-run, command economies in the region.

The economies of the four countries grew rapidly. But in the early 1990s, Japan experienced money problems and so turned away from further investment. To attract money from other sources, the four countries increased the interest rate that they paid on loans. As a result, private investors from Europe and the United States began to invest in the region. Low labour costs allowed the production of inexpensive exports that created enormous profits for company owners and investors. The economic prosperity lasted seven years.

In 1997 a number of banks in the four countries failed; at the same time, a number of manufacturers went bankrupt. Afraid of losing their money, foreign companies quickly withdrew their investments. As a result, even more factories closed, and millions of Asians became unemployed. The governments of the four countries were now faced with the repayment of loans to other nations, leaving them with little money to help the unemployed. Since the unemployed cannot afford to buy much, trade dropped significantly, leading to even more unemployment. Now, at the beginning of the 21st century, millions are still hungry and homeless. There are no schools for many students, and the lack of health services has led to the spread of many diseases. Unhappy and desperate people may yet rise up against their own governments.

The events in Asia have also affected many other parts of the world. Fewer factories in Asia meant less importing of raw materials. This hit the Canadian economy especially hard, because such a large part of our exports consists of raw materials, most of which go to Asia. When the demand for any commodity drops, its value drops too, because there is so much available for sale. Even if we could sell our raw materials to some other nation, we would get less money for them.

The value of money in the Asian countries has dropped. This results in Asian exports being even cheaper to buy in Canada than they used to be. For example, a sweatshirt made in South Korea is much less expensive than one made in

Canada. It is therefore very attractive to a Canadian buyer. A Canadian sweatshirt manufacturer may go out of business because it cannot compete with the cheaper imports. This leads to job losses in Canada.

The effects of the Asian flu show how critical global trade is to the well-being of billions of people. The act of investing in another economy, or withdrawing that investment, can have incredibly far-flung results.

Discover For Yourself

1. Arrange the events below in order from first to last. Then draw a flow chart with arrows linking the events. (You may find that some parts of your flow chart go in circles.) Give your flow chart a suitable title.
   a) Vietnam War ends
   b) Canadian economy declines
   c) Japan invests in Southeast Asia
   d) world commodity prices drop
   e) Asian factories close
   f) Canadian factories close
   g) Japanese economy declines
   h) USA and Europe invest in Asia
   i) Asian economies grow
   j) bank failures and bankruptcies in Asia
   k) investors withdraw money
   l) Asians become very poor
   m) Asians buy fewer imports
2. Explain three ways in which the Canadian economy was hurt by the Asian flu.

Figure 10
Events in Indonesia resulted in this gathering of more than 100 000 people in July 1998 to pray for relief from poverty.

Summary

In this chapter you have learned about the trade links that Canada has with the rest of the world. You have seen that trading agreements play an important role in trading patterns around the world. You have also discovered how we can be affected by economic conditions on the other side of the world.

Reviewing Your Discoveries

1. How has international trade changed over the past 100 years?
2. What is unusual about Canada's main exports? How can Canada's dependence on this type of export hurt Canada when other nations have economic problems?
3. What is the name of the major trading agreement to which Canada belongs? Why did Canada enter into this agreement?

Using Your Discoveries

1. Record your interaction with the rest of the world for a 24-hour period. To do this,
   a) Make a copy of the organizer in Figure 11 and fill it in as the 24 hours progress. (Some examples are provided, but don't copy them down. Also, don't forget to find out about the origins of any TV programs you watch or music that you listen to.)
   b) On a world map, label and shade in the nations of the world with which you interacted during the 24-hour period.
   c) In pairs or small groups, share your findings.
   d) Choose one of the products to research as a group. Find out
      - the company that manufactures it
      - how it's produced
      - who benefits from its manufacture (the nation where it's produced, the company that manufactures it, or the workers who produce it?)

Time	Activity	Links to the Global Economy
7:30 a.m.	washed and dressed	towel from India, T-shirt from Egypt, blue jeans from USA, shoes from Pakistan
8:00 a.m.	ate breakfast	orange juice from USA, drinking glass from France, plates from China

Figure 11
Make a chart like this to record your activities.

Chapter 6

The Local Economy

Key terms

ripple effect

kimberlite pipe

Impact Benefits
Agreement

ripple effect—a chain of
effects or events that results
from an initial event. The chain
is like the ripples caused by
throwing a stone into a pond.

In this chapter we focus on how a single industry can affect a
local economy. The information and activities will help you

▶ identify interrelationships among industries
▶ describe the impact of an industry on a region's economy
▶ evaluate the possible impact of a new industry in your
  school neighbourhood.

"Ripple Effects"

Earlier in this unit, you read about the different economies of
three Ontario communities (see pages 149 to 150). You learned
that both business and government in most communities work
toward a *diversified* economy.

What are the benefits of a diversified local economy? One is
that it saves the community from being dependent on a single
resource. A second involves the ripple effects businesses cause.
The wider the variety of businesses, the wider ranging these
ripple effects will be. Here is just one example.

When the McNeil Consumer Products Company located in
Guelph in 1978, it affected

▶ *building and equipment industries* (needed to construct and
  expand the company building and its machines)
▶ *infrastructure industries* (needed to build and upgrade the roads,
  electricity and water lines, garbage disposal services and sewage
  pipes in the industrial park where McNeil is located)
▶ *housing, furniture, appliance, supply and service industries* (needed
  to supply goods and services for McNeil's 375 employees)

▸ *governments* (the McNeil company and its employees pay taxes to governments at the municipal, provincial and federal levels)

▸ *the local community* (the local community benefits from the taxes paid to the government, and the environment is affected by the company's manufacturing plant).

Figure 1
When the McNeil Consumer Products Company located in Guelph, it started a ripple effect on the local economy.

Old and New Industries

The impact that an industry has in an area depends not only on what the industry is, but also how long it has been there. To illustrate this, we will be studying two very different industries. The first (pages 167 to 170) has been established for a long time. The second (pages 172 to 175) is just starting but is likely to have a large impact on Canada's north.

Case Study
Lester B. Pearson International Airport

Lester B. Pearson International Airport has a tremendous influence on the Greater Toronto Area. It is Canada's biggest and busiest airport, handling over 26 million passengers and 396 000 aircraft movements in 1997. The airport's number of aircraft movements make Pearson the 26th most-busy airport in the world.

The airport, first named Malton Airport, opened in 1938. It was built on farmland purchased by the federal government. In 1962, 1200 ha more farmland was purchased to allow for expansion. Terminal One was opened in 1964, Terminal Two in 1973 and Terminal Three in 1991. Terminals One and Two will be replaced with a single new terminal building, which is expected to be completed in 2006. All the while, new and improved runways and other facilities will be added.

With a large number of employees, travellers, and aircraft to manage, the airport affects the economy of the region in many ways. These ways are summarized in Figure 2 on pages 168 and 169.

Inputs

- **Electricity**
  - From the Ontario Grid
  - For running lighting, equipment, and heating/cooling systems

- **Water Supply**
  - Supplied by Peel County and Toronto
  - For domestic supply and firefighting

- **Aircraft Fuel**
  - From storage tanks connected by underground pipes to outlets

- **Natural Gas**
  - For central heating and cooling of buildings

- **Food**
  - For restaurants and planes

Public Safety
- Police
- Security
- Emergency service

Government Agen
- Canadian: Custom Excise and Taxatio Agriculture Canad Transport Canada, Health and Welfar Employment and Immigration, Can. Post, Environmen Canada
- US: Customs, Immigration and Naturalization

Other Services
- 3 Nurseries
- 3 Medical clinics
- 3 Interfaith centres
- Information booths
- Multilingual guides
- Radio station
- Stores
- Restaurants

Figure 2
Ways in Which Pearson International Airport Interacts with the Region

On the Runways
- 52 Air carriers
- Wildlife control (humane trapping and scaring techniques keep animals and birds off runways)

Transport
- Shuttle buses (run every 10 minutes between 6 a.m. and midnight between terminals)
- Land travel companies

Parking and Storage
- 11 000 Parking spaces
- 8 Cargo buildings

Outputs

► **Wastewater**
- From runoff, terminal buildings and aircraft
- Goes to Peel and Toronto treatment facilities

► **Solid Waste**
- From terminal buildings and aircraft
- Goes to landfill sites

The airport employs about 57 000 people. In terms of the region's population, this means that *one in every 50 workers* in the Greater Toronto Area has a job at the airport! *Another 32 000 people* work at the outside firms that supply the airport with everything from toilet paper to food for the concessions. The spending of these employees supports *a further 24 000 people*. Adding all these statistics together, we can say that the airport

▸ supports *one in every 25 workers* in the Toronto area

▸ generates over $3 billion per year in personal income

▸ generates over $2 billion per year in taxes

▸ generates nearly $12 billion per year in business revenue

By 2005 the airport is expected to handle over 33 million passengers and generate close to 150 000 jobs.

Discover ☼ With Maps

1. Copy the map in Figure 3 that shows the municipalities around Pearson Airport. Outline each municipality.
   a) Choose a small symbol to represent *one per cent* of those who work at the airport.
   b) For each municipality, show what percentage of Pearson's labour force lives there. Do this by drawing in the correct number of the symbol you chose in a). Figure 4 gives you the percentage information you need. Put the "Other" symbols around the map outside the municipalities.

2. Rank the municipalities from highest to lowest number of airport employees living there.
   a) What is the relationship between the number of employees and distance from the airport?
   b) How would you explain this relationship?

WEB LINK

To find out more about Pearson International Airport, look up http://www.gtaa.com

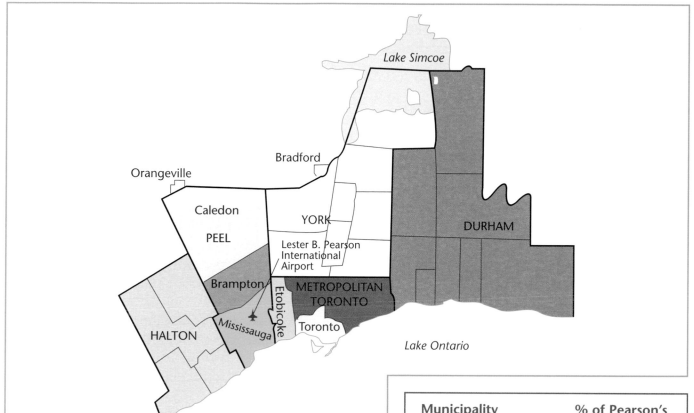

Figure 3
Selected Municipalities Close to Pearson International Airport

3. List and briefly describe 10 ripple effects that result from the building of the airport. Form small groups and share your lists. Make up a group list of the 10 ripple effects you agree are the most important.

4. When the airport was first built west of Toronto, it was surrounded by open countryside. Why do you think this location was chosen? What could the planners foresee?

Figure 4
Percentage of Pearson's Labour Force from Each Municipality

Municipality	% of Pearson's Labour Force
Toronto	14.7
Etobicoke	7.8
Other Metro Toronto	8.1
Mississauga	25.3
Brampton	15.0
Caledon	1.4
Halton	6.8
York	3.9
Durham	2.5
Other	14.5

Case Study   *The Ekati™ Diamond Mine*

A new and exciting kind of mining has recently started in the Northwest Territories: the search for diamonds! Diamonds are first formed beneath the Earth's crust at a depth of about 120 km. When a volcano erupts, these diamonds may travel upward in **kimberlite pipes**. Kimberlite pipes containing gems are very rare—by 1995, only 15 major diamond mines were in production worldwide.

In 1991 about 300 km northeast of Yellowknife, two geologists discovered kimberlite rocks likely to contain diamonds. By 1993 an exploration camp had been built to test the rocks for diamond content. Once geologists confirmed that there were enough diamonds present to make mining profitable, about $700 million was spent preparing for mining operations. This included the following:

- buying heavy mining equipment
- preparing the open pit mine site
- building dormitories for construction and mine workers
- building a power plant and a mill for separating diamonds from the kimberlite
- constructing an air strip
- negotiating **impact benefits agreements**

- interviewing potential employees
- choosing suppliers (e.g., food businesses supplied 100 800 L of milk, 50 400 kg of chicken, 810 000 sandwiches, 324 000 eggs, 30 kg of bacon and 15 192 kg of coffee during the building of the mine. Other businesses supplied 22 000 m³ of concrete and 78 280 kg of steel.)

The Ekati™ Diamond Mine was officially opened in October 1998. It is a joint venture of BHP Diamonds Inc. (51 per cent), Dia Met Minerals Ltd. (29 per cent) and geologists Charles E. Fipke and Dr. Stewart L. Blusson (10 per cent each).

The first rough diamonds, produced in October and November 1998, were sold in January 1999. Customers in Antwerp, Belgium, the world centre of the diamond trade, bought the diamonds for US$8.5 million. Production quadrupled by March 1999.

Spinoff industries are beginning to develop as a ripple effect of the mine. In January 1999, one local manufacturing company opened a processing plant in Yellowknife. This company buys some of the rough diamonds from the Ekati™ mine and cuts and polishes them for the gem trade. Each diamond that the company

---

kimberlite pipe—a roughly cylindrical plug of a rare rock called "kimberlite." The pipe-shaped plug is formed by cooling magma and may contain diamonds.

 impact benefits agreement—a voluntary agreement signed by a mining company and a local Aboriginal group. The mining company agrees to minimize the mine's impact on ecosystems and may guarantee employment to local residents and funding for environmental research.

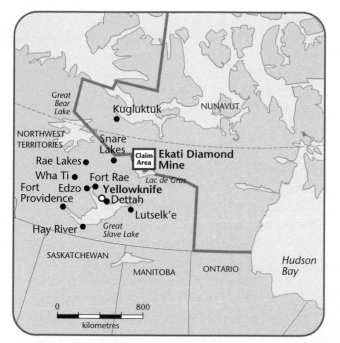

processes in Yellowknife will be laser-engraved with a polar bear logo to show its place of origin. Also in January 1999, a sorting-and-valuation centre was built at Yellowknife Airport, employing about 14 people. Diamonds from the mine are flown there and their value is estimated. On the basis of their value, royalties are paid to the government. Other industries that are expected to develop are gem cutting, jewellery making, and the making of industrial equipment such as diamond drills. For a three-year period starting in mid 1999, 35 per cent of the rough diamonds from the mine will be sold to De Beers Centenary, a well-established worldwide diamond dealer.

Figure 5
Ekati™ Diamond Mine Claim Area

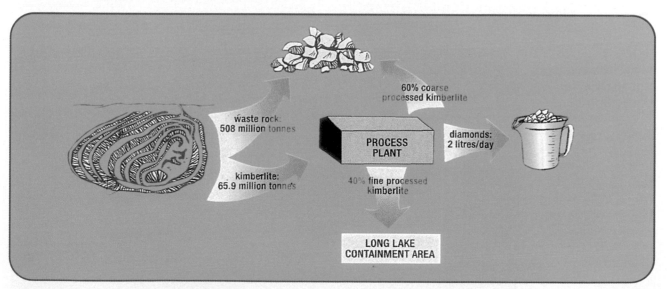

Figure 6
The by-products of diamond recovery are crushed rock and water. Coarse crushed rock is stored with waste rock. Fine crushed rock and water is stored in the Long Lake Containment Area.

The mine will employ about 600 people at high wages. This is a very significant number, since unemployment rates in the Northwest Territories and Nunavut are about 50 per cent higher than in southern Canada, and the population is growing quickly. The mine agreed to hire at least 62 per cent northern residents, at least half of whom are Aboriginal. In reality, it has hired 79 per cent northerners, just over half of whom are Aboriginal. The area from which workers come consists mainly of widely spaced small communities. Wekweti (Snare Lakes), a community of 135 people, is the closest and is 180 km west of the mine. Kugluktuk (Coppermine), with a population of 1201, is probably the furthest community from which workers will be drawn. The mine supplies free air transportation to the mine site for Northwest Territories workers from Yellowknife. Most workers come for two weeks at a time, working 12-hour shifts each day, then return to their communities for two weeks. Some employees work for four days and return home for three. While at the mine site each worker is fed and lives at no charge in one of 375 rooms in five three-storey dormitory blocks. The facilities in the residences include a phone in each bedroom, an exercise room, a games room, a racquetball/squash court, a gymnasium with overhead running track, a dining area and a laundry.

The mine is expected to bring in $14.3 billion in taxes over 25 years. Some of this will return to the area directly or indirectly, as the federal government spends money throughout Canada. The federal government will also award money to the Government of Northwest Territories to pay for infrastructure development, environmental monitoring programs and social services.

Figure 7
Notice the raised tunnel, called an "Arctic Corridor," through which workers travel between the mining facilities and the dormitory blocks. Why is this passageway above ground? (Hint: see page 173.) Why is it enclosed?

Ekati™ Diamond Mine has agreed to buy 70 per cent of its goods and services from the northern businesses, with priority given to those that are owned and operated by Aboriginal people. Goods include food, clothing, equipment, and supplies. Services include computer support, contractor services and travel. The Ekati™ mine will also provide scholarships to northern students.

One major concern that most people have when a mine opens is the damage that it will do to the environment. The mine has been built in a tundra environment with underlying permafrost (permanently frozen subsoil). The land is easily damaged by heavy equipment, especially in the summer when the surface layers of permafrost thaw. The mine operators have agreed to try to alter the land as little as possible. They have also agreed to restore the environment after they have finished mining. The environmental staff includes 10 experts who study the environment, monitor it and advise how to reduce impacts from mine operations. Hunters and naturalists in the area are especially pleased that so much care is being taken to protect the environment.

 Discover *For Yourself*

1. Create an organizer with two columns. In the first column, list all the benefits that will probably come to this northern area as a result of the mine opening. In the second column, list the problems that might result from the mine's opening.
2. Write a page comparing the economic impacts of Ekati™ Mine and the Pearson International Airport. Include information about employment, ripple effects and the environment.

SCIENCE LINK

WEB LINK

To find out more about the Ekati mine and its ripple effects, look up http://www.bhp.com.au/minerals/diamonds/index.htm

Figure 8
Caribou and other wildlife make their home in the area of the Ekati™ mine. How can an environment's wildlife be affected by economic activity? How can industries control these effects?

Summary

In this chapter you have discovered that an industry can have a tremendous impact on a local economy. You have seen one example of such impact by a long-established operation (the Pearson International Airport). You have seen a second example of such impact by a new industry (the Ekati™ Diamond Mine).

Reviewing Your Discoveries

1. Give an example of
   a) a direct advantage of an industry locating in a community.
   b) an indirect advantage of an industry locating in a community.
   c) a disadvantage of a heavy industry locating near a city.
   d) a disadvantage of a mine locating in a tundra environment.

Using Your Discoveries

1. In small groups, decide on a new business that you think could prosper in your school neighbourhood. Possibilities include: a small restaurant; a fast-food restaurant; a video store; a clothing store; a car manufacturing plant; a gas station; a plumbing parts manufacturing plant; a local newspaper; a computer and electronics salesroom; or any other businesses you choose.
   a) Choose a precise location for your business.
   b) Prepare a presentation to your classmates that explains the benefits your business would bring to the region. Each presentation should include a map, two flowcharts (one showing the business operation and a second showing its ripple effects) and statistics on employment.
   c) Give your presentations.
   d) Analyze each other's presentations.
      – How realistic are the benefits?
      – How could the benefits by maximized?
      – What disadvantages could result?
      – How could the disadvantages be minimized?

Geography Workshop

This unit has presented information on many aspects of economics, including these five:

▶ economic activity (buying, selling, and exchanging goods and services)
▶ economic resources (land, labour, capital, entrepreneurial ability)
▶ economic systems (subsistence, traditional, command, market, mixed)
▶ economic sectors (primary or resource, secondary or manufacturing, tertiary or service)
▶ Canada's economic features (industrialized with strong secondary and tertiary sectors, mixed, globally linked through trading agreements such as NAFTA)

In this **culminating activity**, you will use your knowledge of economics to analyze articles about the Canadian and the global economy.

An Example

Read the following article and the analysis of it.

Brave New Brands

Open the fridge and enter the brave new world of global branding. Thirsty for a sip of Canada Dry? The last time a Canadian company produced it was in 1931. The label is now owned by the British soft drink and candy giant Cadbury Schweppes PLC, which sells it in 90 countries. In the den, the RCA trademark emblazoned on that big-screen TV belongs to Thomson C.S.F. of France. The car in the garage might be resting on Firestone tires, a famous U.S. brand now owned by Bridgestone Corp. of Japan. And if the vehic[le is] a Chrysler, even that venerable American nam[e] will soon pass into the hands of Daimler-Ben[z] AG of Germany.

Big-name brands are in demand, and com[m]panies are willing to pay huge amounts for t[he] instant recognition and market access they afford. With the spread of television through[out] the developing world, brand-consciousness h[as] become a global phenomenon. In the proce[ss,] the premium that companies place on high-profile products is rising.

Above all, brands make good business se[nse,] says Niraj Dawar, a professor of marketing at the University of Western Ontario in London[.] "The cost of building a brand today is very high, so the value of existing brands become[s]

Most Important Points

Brand names help to sell products because consumers recognize brand names.

This is happening all over the world, even in developing countries, because many people watch TV commercials.

Companies will spend a lot of money to buy a brand name.

The costs of developing and advertising products are very high, so it helps to get a larger market to buy the products. Brand names will get that larger market.

Links to Economics Unit

Chapter 1, page 106—This article shows the relationships between consumers, producers, advertisers and market researchers.

Chapter 3, page 134—The big companies mentioned in the article are all from the world's most industrialized nations (the G-7 nations): Britain (UK), France, Japan, Germany, and the United States.

Chapter 4, page 145—This article shows market economies at work. The companies have let the market determine the names of their products.

ater and greater." The relentless push multinationals to promote their brands bally is driven partly by a desire to spread duct development and advertising costs r a large market, says Dawar.

Where a local brand is considered weak, npanies will replace it with a stronger ltinational name. In Canada last year, for example, U.S.-based PepsiCo eliminated the Hostess potato chip brand in favour of its Frito-Lay moniker.

—**John Schofield**
Abridged from Maclean's,
May 18, 1998, p. 32

Your Job

Find two recent articles in a newspaper, news magazine or on the Internet. One should be about the economy within Canada. The other should be about economic activity involving Canada and one or more other countries. After finding and reading each article, follow these three steps:

1. Clip or make a copy of the article. Trim and mount it and add the date and name of the source publication.
2. Summarize the article in your own words, using point form as shown in the example.
3. Show how the contents of the article are linked to specific things you learned in this unit.

Unit Three

Discovering Human Movement

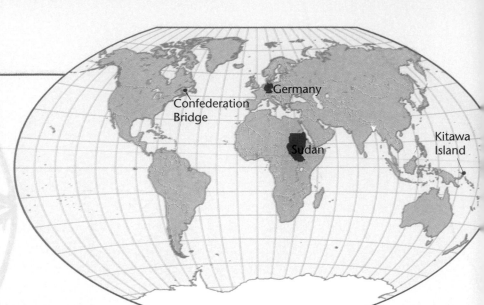

In Unit One, one of the human patterns you studied involved settlements—the small towns or large cities that people have built to live in. People are constantly moving within and between these settlements. Some of them are moving by choice, while others are forced to relocate. Geographers want to understand this movement. They want to learn about where people move from and to, how they make their journeys and why they do so.

This unit will help you discover patterns in human movement. You will see which parts of the world people want to move away from and which parts they want to move to. You will learn how this movement affects a country such as Canada, which has become multicultural because of movement. You will also study the effects of technology on movement and travel. While you do so, you will "visit" places and regions around the world, including those highlighted in the world map on this page.

Chapter 1

The Spread of Culture

Key terms

culture
diffusion
multicultural society

culture—the behaviour that people learn, made up of their belief systems, languages, social patterns, political systems, organizations, food and clothing customs, and use of buildings, tools, and machines.

In this chapter we focus on how cultural features spread. The information and activities will help you

▸ identify patterns in the locations and spread of various cultures
▸ show an understanding of how cultures are affected by migration
▸ show an understanding of the effects that migration has had on Canadian culture and your sense of identity.

North American Culture

One way to understand **culture** is to look at examples of our own. See what features of North American culture you can identify in the following description.

> At 7:30 Angela Marshall was jolted from a deep sleep by her radio alarm. News and weather were followed by rock music. She struggled out of bed and stumbled into the bathroom. She stepped into the warm shower and in a few minutes felt better. She blow-dried her hair, put on some makeup and dressed in blue jeans, T-shirt and a heavy sweater her mother had given her for Christmas. There had been a frost but the forecast was for a warm afternoon.
>
> At the breakfast table, Angela poured herself a glass of orange juice from a carton and a bowl of sweetened cereal.

Her step-father was checking the e-mails and her mother was setting the automatic timer on the oven so the frozen lasagna would be cooked by suppertime. Angela's parents would be leaving, each in their own car, to drive to work in five minutes. Her step-father would take Angela's baby brother to the sitter's on his way to work and her mom would pick him up on her way home from work.

Angela left the house at 8:30, making sure that the front door locked behind her. Today there was a math test, but she was looking forward to the baseball game after school.

The activities described in Angela's story are part of North American culture. The North American culture includes

▸ a variety of family types (some two-parent, some single-parent, and other types)

▸ many families where both adults work away from the home

▸ English, French or Spanish as the usual languages spoken at home

Figure 1
What clues does this bedroom give about North American culture?

▸ teenagers who like popular music, blue jeans, and "hip" clothing
▸ the use of much electrical equipment and highly processed convenience foods
▸ a high regard for hygiene and the use of much water
▸ the use of makeup by teenage girls and women
▸ dating by young people.

The North American culture is a normal way of life for about 5 per cent of the world's people.

 **Discover For Yourself**

1. Divide into small groups to discuss these questions about North American culture.
   a) How can you tell from Angela's story that a fair amount of wealth is typical in North American culture? Find four examples to support your answer.
   b) Imagine that the North American culture changed in the following way: *In families with more than one adult and any children under 12 years of age, it is the custom that one adult, usually the father, will stay home to care for the home and children.* How would this affect the way that many people live?

The Development of Culture

Every culture in today's world had its beginnings in one or more centres of the ancient world, shown in Figure 2. In these centres, farmers learned to produce enough food to feed large groups of people. As a result, other people in the centre had time to develop the language, writing, religion, art, and technology of their group. Eventually, civilizations evolved and prospered in each centre.

As travellers and traders moved between different centres, they exchanged ideas with those they met. This led to the spread, or **diffusion**, of certain features of each culture to surrounding areas.

Today's North American culture is a result of centuries of such diffusion. Think back to Angela Marshall. The design for the bed she wakes up in originated in Asia and was modified in Northern

diffusion—the spread of a new feature from a centre or centres. For example, the use of chili peppers for cooking and eating began in Central America about 5000 years ago. Explorers and settlers caused it to spread (diffuse) to Europe. It then diffused to South Asia with Portuguese settlers, and is now a basic ingredient in much South Asian food.

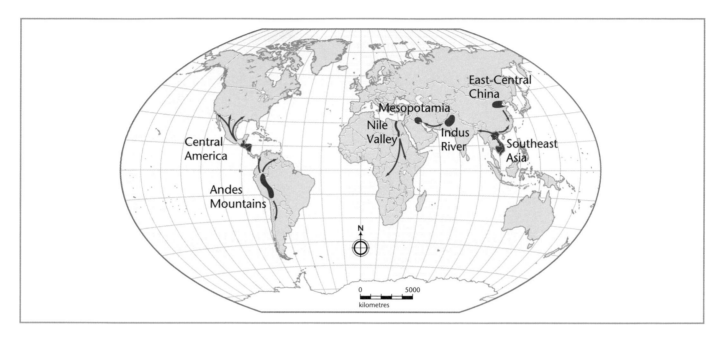

Figure 2
Major Centres Where Different Cultures Developed

Europe. Her sheets are made of cotton, which was first developed in India. India is also where the pajamas she sleeps in were invented. The moccasins she wears to walk to the bathroom were invented by North American Aboriginal peoples of the Eastern Woodlands. The soap she washes with was invented in ancient Europe. The cereal bowl she uses is made from pottery invented in China. The steel of her cutlery was first made in southern India, and the fork and spoon she eats with were invented in Italy.

Diffusion Diagrams

Diffusion diagrams show how a feature spreads from one centre to other places. A simple way of showing diffusion is to draw arrows on a map, as in Figure 2.

Geographers who study culture are interested in the diffusion of cultural features. Figure 3 (on page 186) shows the diffusion of major languages around the world. Figure 4 shows the diffusion of major religions around the world.

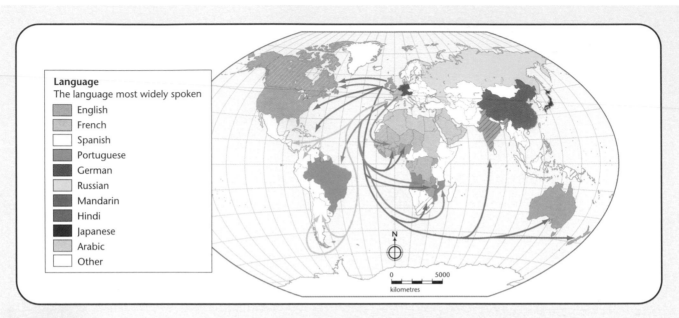

Figure 3
The Diffusion of Major Languages Around the World

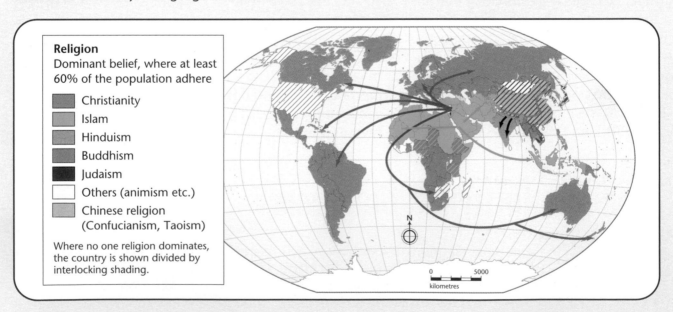

Figure 4
The Diffusion of Major Religions Around the World

  **With Maps**

1. Look at Figure 3.
    a) Which continent has had a great influence on the spread of languages throughout the world? Give examples of the spread of four different languages to support your statement.
    b) Suggest a reason why the continent you identified in a) was so influential in the spread of language.
    c) Why do you think the nations that make up Eastern Europe did not spread their languages around the world?
2. Look at Figure 4.
    a) Which religion is the most widespread as the dominant belief? List the regions to which this religion has spread.
    b) Which religion is the least widespread as the dominant belief? In which country is this religion dominant?
3. Draw a compound bar graph to show the data in Figure 5. Which continent has large numbers of people in each of the four religions?

Turn to page 264 to learn how to draw compound bar graphs.

Religion	Africa	Asia	Europe	Latin America	North America	Oceania	World
Christianity	348.2	306.8	551.9	448.0	249.3	23.8	1 928.0
Islam	300.3	760.2	32.0	1.3	5.5	0.4	1 099.7
Hinduism	1.5	775.3	1.5	0.7	1.2	0.3	780.5
Buddhism	0.0	320.7	1.5	0.6	0.9	0.2	323.9

Figure 5
Number of People (in Millions) Belonging to Four Major Religious Groups

A Sense of Identity

Each of us forms a sense of identity from the combination of language, religion, and other features that make up our cultural background. This sense of identity is very important. It anchors us to the physical world and gives us a base from which we can observe, take our place in, and change the world. This can only happen when we are accepted for who we are and accept others likewise.

multicultural society— a country or part of a country where large proportions of the population are from different cultural backgrounds. In such a society, people are encouraged or allowed to maintain their cultural traditions.

In Canada, laws have been passed and institutions have been funded to help make sure people from all cultures are accepted. This makes Canada a **multicultural society**. Living in such a society allows us to enjoy many of the foods, traditions and other cultural features brought here from other parts of the world.

Many other nations also have citizens from different cultures. Others have populations that are almost all of one cultural group. Figure 6 shows examples of how these differences have arisen.

The differences among cultural groups living in one country sometimes result in tensions and even war. This is especially true in nations where one cultural group has much more power than

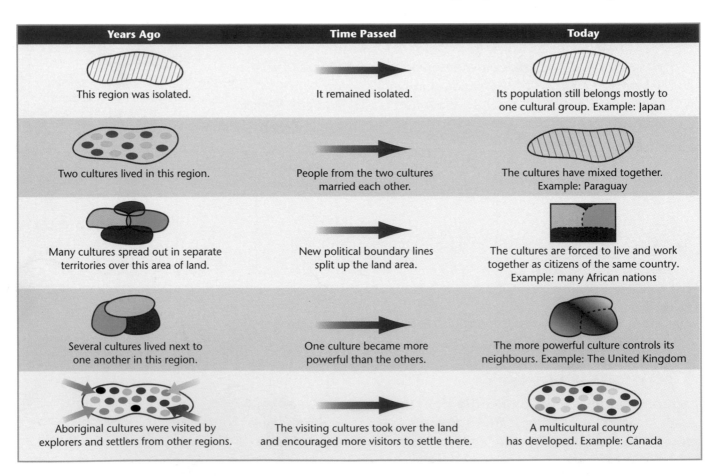

Years Ago	Time Passed	Today
This region was isolated.	It remained isolated.	Its population still belongs mostly to one cultural group. Example: Japan
Two cultures lived in this region.	People from the two cultures married each other.	The cultures have mixed together. Example: Paraguay
Many cultures spread out in separate territories over this area of land.	New political boundary lines split up the land area.	The cultures are forced to live and work together as citizens of the same country. Example: many African nations
Several cultures lived next to one another in this region.	One culture became more powerful than the others.	The more powerful culture controls its neighbours. Example: The United Kingdom
Aboriginal cultures were visited by explorers and settlers from other regions.	The visiting cultures took over the land and encouraged more visitors to settle there.	A multicultural country has developed. Example: Canada

Figure 6
In these examples of cultural mixes around the world, each colour represents a different culture or sub-culture.

the others. On the other hand, those who live in a country made up of one culture often have an unnecessary fear of other cultures. They have not had the chance to be friends with those who have different cultural backgrounds.

Case Study   *A Disappearing Culture*

As people interact with and change the Earth's environment, certain species of animals and plants can no longer survive. As a result, Earth's ecosystems are much less varied, or diverse, than they once were. The same can be said about *cultural diversity*. Some cultures cannot survive in an environment that is overrun by North American and European influences.

In Canada, many members of Aboriginal cultures have stood up against these influences. They have asked for their traditions and lands to be recognized. Their successes are impressive, usually coming after decades of struggle. The same cannot be said right now for members of the Kitawan culture in the South Pacific. This culture is in danger of disappearing forever.

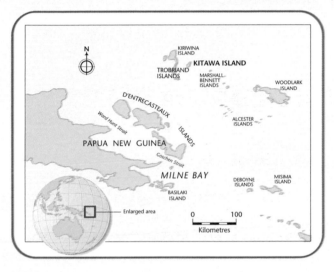

Figure 7
Kitawa and the Surrounding Area

Ancient Traditions

On a few hard-to-reach islands in the South Pacific, several cultural groups have lived in isolation for 10 000 to 15 000 years. They include the Kitawans on the island of Kitawa. The Kitawans, living in three villages on the island, have developed a unique set of traditions that can be seen in their art, carving and poetry. These traditions have been passed on by word of mouth because the culture has no writing system.

One key feature of Kitawan culture involves the making of 20-metre-long ceremonial canoes.

An entire village is involved in creating each canoe. A canoe carries 40 men from island to island twice each year. During these voyages, the Kitawans exchange gifts with a number of partners on other islands in a carefully designed pattern. In the course of a year, all the travellers will have visited each island, and all gifts will have passed through the hands of all the partners.

The first step in building the huge canoe needed for this ceremony is to choose a suitable tree. After the tree is cut down, it is moved many kilometres from the forest to the beach using vines,

Figure 8
A Kitawan Man and His Son

ropes, and ramps. It never touches the ground until it reaches the beach. Then the trunk is hollowed out. A master carver adds a beautifully designed prow, or front board, to the trunk. The carvers are revered by other villagers. It is considered an honour to be accepted for training as a carver.

Carving, like all traditional activities, involves passing down skills, rules, and designs from one generation to the next. The passing down of knowledge is carefully organized, with each family in charge of a different type of knowledge. For example, one family is responsible for passing on dance knowledge, while another is in charge of music, and yet another is expert in body decoration.

Traditions Under Threat

Kitawa can only be reached by boat, and frequent storms and other rough water conditions make the trip very hazardous. Only a handful of Westerners have ever been to the island. But the other islands in the area are another story.

Contact with Western culture has brought motorboats, canned food, Western tools and clothing to the islands around Kitawa. Kitawans visiting the surrounding islands have seen traditions disappearing. Many Kitawans now question the value of maintaining their own culture. In 1990, they stopped holding the traditional harvest dance. No canoes have been built for several years. Many islanders trade their agricultural products for manufactured goods. Items made of plastic and aluminum are high on the list of goods they want. Western shoes, even though they hurt the hard-soled feet of the islanders, are also in demand. Since the economy is subsistence, this trading away of food means there is not enough to eat in the villages. The elderly and infirm are hit especially hard by food shortages.

When independence came to Papua New Guinea in 1975, the people chose English to be its first language and Christianity as its major religion. Many Christian churches discourage the ancestral traditions of the islanders. Some Kitawan children have been sent to school on other islands where they get a taste for Western values and possessions. Very few make it into high school, so they return home. But their schools have taught them to disapprove of the traditions at home, and they have not learned how to fish, sail, or grow food. Without these vital skills needed to survive, they are unhappy in their society. As a result, the culture is steadily disappearing.

Discover For Yourself

1. Imagine you lived on Kitawa before there was any contact with the outside world. What would you have pictured your adult life to be like? Draw a picture to illustrate your description.
2. Divide into small groups to discuss these questions.
   a) Why is a culture with no writing much more likely to die out than one that has writing?
   b) What kinds of skills and traditions will probably die out as a result of Western influence on Kitawa?
3. Write and perform a live skit or record a skit on video. In the skit, hold a conversation among a small group of young teenagers who have returned home to Kitawa for school vacation after their first year at a school on a distant island.

Summary

In this chapter you have discovered the different features that make up a culture. You have learned about how cultural features such as language and religion have spread. You have also thought about how culture gives us a sense of identity. Finally, you have read about a culture that is likely to die out in the near future.

Reviewing Your Discoveries

1. List four things that are typical of North American culture.
2. List three ways in which cultures may differ in other parts of the world.
3. Your identity is made up of many factors.
   a) List two major factors that contribute to your identity.
   b) List and briefly explain three other influences on your character and outlook on life, school, and family.
   c) Share your findings with your classmates.

Using Your Discoveries

1. In small groups, research one of the world's **indigenous peoples**. Some possibilities outside North America are: the Ainu

indigenous peoples—cultural groups who lived in an area from early times before the arrival of colonists.

(Japan), the Basque (Europe and Russia), the Carib (Caribbean islands), the Maori (New Zealand), the Penan (Sarawak, Malaysia), the Taino (Central America), the Tamil (southern India and Sri Lanka), the Yanomami (Venezuela and Brazil). For the culture you chose, discover
- the traditional way of life (as shown in food, shelter, clothing, dances, art, music, celebrations and so on)
- what, if any, threats there are to their existence
- what, if anything, is being done to help them
- what will happen to the people if their culture weakens or disappears
- what the world will lose if the culture disappears.

Present your findings in one of the following forms:
- a written report
- a poster with illustrations and written paragraphs
- an oral presentation.

Be sure to include in your findings the sources of information you used.

Figure 9
These Penan tribeswomen are making colourful rattan bangles and bags in a longhouse in the Sarawak interior.

Chapter 2

Understanding Movement

In this chapter we focus on types of human migration. The information and activities will help you

▸ show an understanding that migration results from decisions people make about their environment
▸ identify factors that influence people to move away from or to another place
▸ identify barriers to migration.

Key terms

migration
refugees
immigration

migration—movement from one area to another.

A Changed Environment

The history of human **migration** goes back a very long time. It goes back well beyond the time of ancient civilizations, when cultural features started to spread through exploration and trade (see Figure 2 on page 185). In fact, it goes back 50 000 years to the beginning of modern human existence. This has been called the time of *Homo sapiens sapiens* or "Cro-Magnon" people.

This first phase of modern existence was also a period of ice ages. As worldwide climate patterns fluctuated great ice sheets advanced and retreated over the Earth's surface. These climate changes affected where different types of vegetation grew and how soil developed. As a result, wild animals moved to find new grazing areas. Predators, along with our early hunting ancestors, moved with them. As the climate changed from wetter to drier and from warmer to colder, farmers too were forced to move to areas with the most suitable climate for their crops.

Since great ice sheets covered the high latitudes, much of the world's water was trapped in the glaciers and sea levels were much lower than today. People could walk or take short boat rides between land areas that are now separated by deep water. Figure 1 shows a series of human movements that occurred during the last Ice Age. They resulted in the spread of humans over most parts of the Earth's land surface.

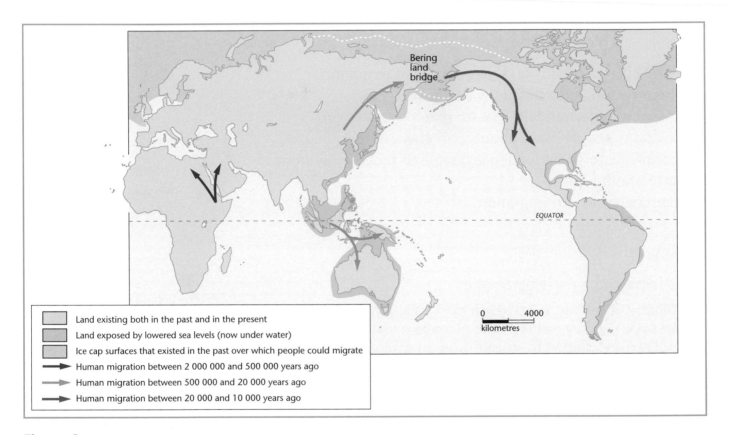

Figure 1
Routes of Human Migration
Between 2 Million and 10 000
Years Ago

Discover *With Maps*

1. Describe three main migration patterns shown by arrows in Figure 1. Include *where the movement was from*, *what direction the movement went in*, and *the destination of the movement* in your answer.
2. How was the environment of the past different from today? How did this different environment help migration?

Environmental Changes Today

There are still millions of people who migrate because of changes to the environment. As the climate changes seasonally, nomadic herders move from one place to another searching for good pasture. This happens in Africa, Asia, and the Arctic. Figure 2 shows an example.

Many migrant agricultural labourers work in fields and orchards in the warmer parts of the United States during the winter. They move northward as the weather warms up in the summer. They help harvest farm products all the way to southern Canada. Over 20 years ago, the Canadian and Jamaican governments signed an agreement in which Canadian farmers pay for the airfare, accommodation and labour of Jamaican migrant workers. The workers travel from Jamaica to Canadian farms in April and stay until October, receiving minimum wage as pay. Twenty-five per cent of this pay goes directly to the Jamaican government, which pays most of it back to the workers when they return home for the winter.

Within your lifetime, climatic changes may cause the migration of millions, even billions, of people from lowland areas. This is because sea levels may be rising as global warming causes glaciers and ice caps to melt. Nobody is sure how much sea levels will rise, but one current estimate suggests a one-metre rise by the year 2050. If this happens, 150 million people worldwide would have to move to higher land. Much of this lowland area is heavily populated and produces a large proportion of the world's crops. A six-metre rise would displace 40 per cent of the world's population. These people would have to find housing and work on higher land.

Figure 2
The Nenets are reindeer herders. Every year they migrate between winter pasture in northern forests to summer pasture in tundra.

Figure 3
Jamaican Migrant Worker in Ontario

Discover With Maps

1. Look at the map in Figure 2 on page 195.
   a) Use the scale to measure the length of the Nenets' yearly migration route.
   b) The lands used by the Nenet and their reindeer hold what may be the largest natural gas reserves in the world. More than a thousand gas industry workers have moved into the area drilling gas wells and building houses, roads, and a railroad. Describe how you think the environment and the Nenet people will have changed by the mid 21st century.
2. Divide into small groups to look at the maps in Figure 5. Brainstorm the effects of a six-metre rise in sea level on Florida's
   a) people
   b) transportation links
   c) agricultural production
   d) tourism

refugees—people who have fled from their own country because of war, natural disaster or persecution based on race, religion, nationality, social group, or political opinion.

War and Famine

Changes in the environment caused by climate are one reason people migrate. War and famine also change environments and cause migration. When the migrants in these circumstances flee to another country, they become **refugees**. Figure 4 below shows the top five sources of refugees as of January 1, 1999.

Country of Origin	Number
Afghanistan	2 648 000
Iraq	631 000
Bosnia	597 000
Somalia	525 000
Burundi	517 000

Figure 4
Highest Ranking Sources of Refugees, January 1999

Source: United Nations/*Maclean's*, August 23, 1999, p. 18.

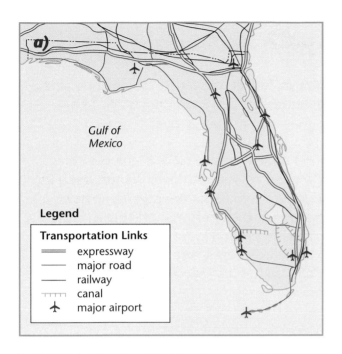

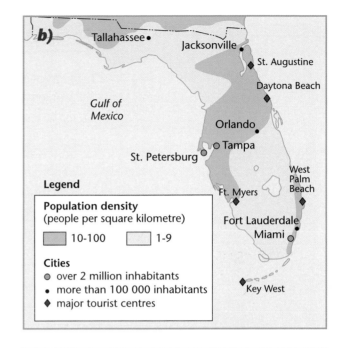

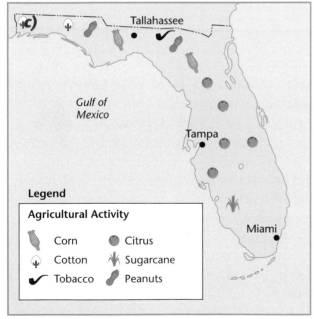

Figure 5

A 6-metre rise in sea level (map (d)) would affect transportation (map (a)), population distribution (map (b)) and agriculture (map (c)).

Discover 🔍 For Yourself

1. Geographers use the term "driven migration" to describe all the movements outlined in the chapter so far. Why is this term appropriate? Include examples in your answer.
2. Look at Figure 4 on page 196.
    a) Make or get a world outline map. Colour in the countries in Figure 4. Use the darkest colour for Afghanistan to show its high number and lighter colours for the other four countries.
    b) Divide into groups of five. Each person in the group should choose a different country from Figure 4 and research its recent history. Make a group list of the reasons that would explain why and how a country becomes a high-ranking source of refugees.

slave—a person who is owned by another and who must do what the owner wishes.

Forced Migration

Forced migration occurs when a powerful group of people forces others to move against their will. The international **slave** trade is an example. It began around the Black Sea centuries ago when men were captured to fight in Egyptian armies and women were captured to be servants in Western Europe.

The slave trade in Africa began in the 1440s when Portuguese sailors started taking Africans to Europe. Between 12 and 30 million Africans were forced to move as slaves to the Americas during the 1700s. Many died on the boats that transported them, which were crowded and unsanitary. When those who had survived reached their destinations, they were sold to landowners who needed labourers for their fields, plantations, and orchards. Some also worked as servants, road builders, and miners. Today, 40 million people in North and South America and the Caribbean are descended from those slaves.

At the end of the 1700s, the British government sent many criminals to Australia. This served two purposes. It rid Great Britain of people they no longer wanted in the country, and it helped to populate the colony of Australia. The forced migrants included Elizabeth Hannell, a young woman convicted of stealing a loaf of bread. She was shipped to Australia and chose to stay there once she had finished her sentence. She eventually had seven children. Her eldest became the first police chief and Lord Mayor of the City of Newcastle, 100 km north of Sydney.

Many people are unaware that Canadians have also been forced to move from their homes by the Canadian government. In the 1950s, 92 Inuit people were forced to move. They had been living in Inukjuak, in northern Quebec, and Pond Inlet on Baffin Island. The government sent them to Craig Harbour and Grise Fiord on Ellesmere Island and to Resolute Bay on Cornwallis Island. The reason for this forced migration was to help Canada claim ownership of the two sparsely populated northern islands. Hunting in the new land was poor, and the migrants were largely neglected as they struggled to survive in a much harsher environment.

During the Second World War, thousands of Canadian men, women, and children who had Japanese ancestors were moved from coastal areas of British Columbia to internment camps (prisons) away from the coast. The government feared that they might help Japan invade North America during the war. But this fear was unfounded.

Africville, a suburb of Halifax, Nova Scotia, had been inhabited by African Canadians since the early 1800s. On a cold night in 1969, bulldozers flattened Africville to make way for a park. The close-knit community, which had been there for eight generations, was broken up, and people were moved into other parts of the city.

Figure 6
Many Canadians with Japanese ancestors were forced to move away from the coast of B.C. during the Second World War.

Figure 7
Africville, Nova Scotia

Discover  *For Yourself*

1. Make a copy of the following organizer and fill it in.

Type of Migration	Description	Example from Over a Century Ago	Example from the 20th Century
Driven			
Forced			

2. Re-read the account of the Inuit people who were moved.
   a) Using an atlas, locate the places mentioned (Craig Harbour is located on the southern tip of Ellesmere Island, just east of Grise Fiord). Draw a simple outline map of that part of Canada and mark the five places. Also label Quebec and the main islands.
   b) In small groups, look at thematic maps from an atlas of the area you drew in your map. Use the information shown in the maps to discuss this question: *Which group of Inuit—those from Inukjuak or those from Pond Inlet—would have had the greatest problems in adapting to their new home?* Give reasons to support your answers.
   c) The Inuit migrants moved in the late autumn. How did this create an additional problem for them?

3. In small groups, research the forced migration of Japanese Canadians or the residents of Africville. After you have learned the details surrounding the migration, hold a role-play discussion showing the different points of view of the people involved.
   a) If you chose the Japanese Canadians, different group members should role play:
      – Prime Minister Mackenzie King
      – a Japanese Canadian taken to an internment camp
      – a Japanese Canadian sent to work as a labourer in Ontario
      – a B.C. resident not of Japanese background
   b) If you chose Africville, group members should role play:
      – a resident of Africville
      – a developer interested in developing Africville land
      – a member of the Halifax City Council in charge of relocating Africville's residents

Voluntary Migration

Many people choose to leave the country of their birth to make a better life for themselves elsewhere. They are looking for a better job, adventure, or escape from the things they do not like in their home country. This type of movement is known as "free migration."

Once one migrant has successfully become established in a foreign country, other family members may wish to follow. This is known as "chain migration." These people enter under a family reunification program. They usually do not have to meet all of the same strict conditions as the original immigrant.

Occasionally, a large number of people move from one part of the world to another. This is called a "mass migration." During the 1800s and 1900s, 45 million people migrated from Europe to North and South America. Their lives in Europe had been difficult, and advertisements encouraging them to move to the Americas were widespread. Many came because family members who had already migrated wrote hopeful letters home, reassuring them of better lives across the Atlantic Ocean.

The United States, Canada, and Australia are the only countries that still allow mass immigration. They are now starting to be more restrictive.

Figure 8
This boatload of immigrants is one of many that arrived in Canada from Europe during the 1950s.

Discover With Maps and Graphs

MATH LINK

1. Divide into groups of five. Each person in a group should choose a different graph from Figure 9 to analyze. To analyze your graph, follow these steps.

    a) On an outline map showing countries of the world, colour in and label the nation represented in your graph. Use an atlas to help you.

    b) On your map, write numbers on the countries from which immigrants to your nation come. Write a number "1" on the country from which the largest proportion of immigrants come. Write a number "2" on the country from which the second-largest proportion of immigrants come. Write a number "3" on the country from which the third-largest proportion of immigrants come.

    c) Look at your maps together and discuss as a group whether the following statement is true or false for each of the countries: *The largest group of immigrants comes from a country in the same continent.*

2. In your groups, discuss these questions about the five graphs in Figure 9.

    a) How do the places of origin of Canadian immigrants differ from those of the United States?

    b) How do the places of origin of immigrants to Australia differ from those going to France?

    c) List three factors that would help nations decide from which countries they should accept large numbers of immigrants. Give examples to back up your factors.

Barriers to Migration

There are four main factors that act as barriers to moving easily from one place to another. The first is *physical*—mountains, deserts and other physical environments cannot support many people and thus keep people from migrating to them. The second is *financial*—the costs of applying for **immigration** to a country, then travelling

immigration—the act of people entering and settling into a country different from their native country.

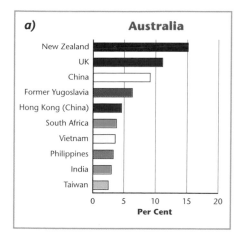

a) Australia

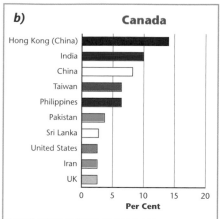

b) Canada

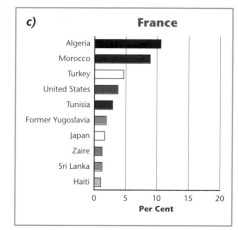

c) France

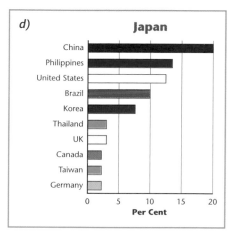

d) Japan

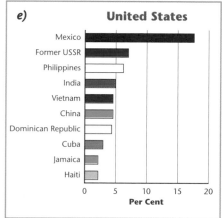

e) United States

Figure 9
Percentages of Immigrants by Country of Origin

to it and setting up a new home there, can prove too high for many. The third is *legal*—applicants for immigration to another country have to meet many requirements set up by the country. The fourth is *emotional*—it is hard for a migrant to leave family, friends, the cultural activities and the places that have become part of his or her life. Immigrants may worry about these factors:

▶ cost of airfare
▶ high altitude in new country that makes breathing problems worse
▶ different language spoken in new country
▶ lack of qualifications required by new country
▶ pain of leaving friends

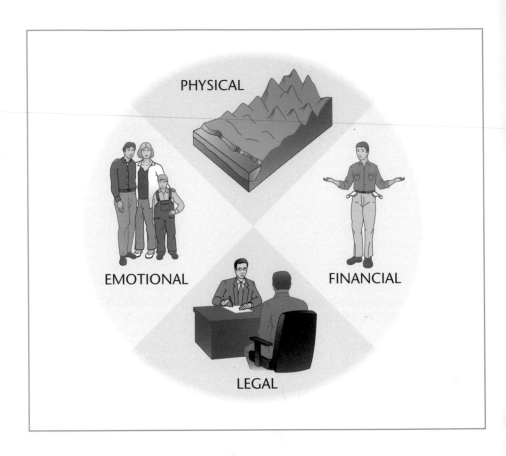

Figure 10
This diagram illustrates the four main types of barriers that immigrants may face.

To immigrate to Canada, a person must first pay an application fee of $500. Other expenses include the following:

- Landing fee: $475
- Immigration lawyers (optional): $250–5000
- Fingerprint checks at police department: $35
- Cost of getting education transcripts
  and other certificates: $40
- Chest X-ray: $30

From a legal point of view, immigrants to Canada must qualify by getting a minimum number of points in areas such as education, training, and job experience. Points are also awarded if Canada has a demand for the immigrant's job skills and if the immigrant has employment already arranged. Knowledge of French and English, age, and personal suitability are also included in the points system.

Case Study

Facing the Barriers

Patricia Abeliamba is 45 years old, with a son Michael who is now 4 years old. For two years, she and her son lived with a Canadian family working in Kenya. She did most of the cooking, cleaning and shopping for the family, and looked after the family's two children. In exchange, she received meals, a room to live in and a small salary. She sent most of this money to the father who looks after their three other children and two grandchildren. Her employer, Maryanne Cleave, was very happy with Patricia's work and asked her to come back to Canada with the family. Patricia obtained the required visa to enter Canada. She also got a work permit so that she could work while she was here.

A few months after the Cleaves, Patricia and Michael had arrived in Canada, Maryanne Cleave received a phone call from Immigration Canada. Maryanne was informed that Patricia's work permit would not be renewed at the end of 12 months. So Patricia would be sent back to Kenya after a year. Patricia would really like to live permanently in Canada but it is unlikely that she would be accepted as an immigrant. Patricia hopes to return to Canada next year with a new work permit. Her family really benefits from the money that she can send home, and living in Canada gives Michael a chance to learn English.

Discover For Yourself

1. Divide into small groups to analyze the case study. Work on these questions:
   a) List and briefly explain four factors in Canada's immigration criteria that might prevent Patricia and Michael from becoming immigrants.
   b) As far as you can tell, could Patricia and her son come to Canada as refugees? Why or why not?
2. Imagine that you are 21 years old and have immigrated to Canada from another country. You are about to leave Pearson International Airport. A cab will take you to an inexpensive motel where you will stay until you have found a job and a place to live.
   a) Describe your emotions in a diary or journal entry.
   b) State three questions that are going through your mind.
   c) List and describe five services that would be especially helpful for you as an immigrant.

LANGUAGE LINK

Summary

In this chapter you have discovered that migration can be driven, forced or voluntary. You have learned why some people are prevented from migrating. You have also been introduced to some of the immigration policies of Canada.

Reviewing Your Discoveries

1. List the causes of driven migration.
2. Describe the circumstances of forced migration.
3. Explain how free migration can lead to chain and mass migration.
4. What are the four types of barriers to migration? Give one example of each.

Using Your Discoveries

1. Research the "Underground Railroad." In a short written report,
   a) explain the reasons why it developed
   b) produce a map to show where it was located
   c) explain the effect it had on Ontario's development.
2. Imagine you are a member of a group of 20 nomadic herding families. Play the "Search for Perfect Pasture" game in Figure 11. When you have finished the game,
   a) list the physical barriers that you encountered
   b) describe one other physical barrier not shown on the map that might have affected your movement.

3. It is possible that at some future time people will leave the Earth to live in space or on another planet. Divide into groups to write or perform a skit about migration to space. Make sure that at least one group chooses each of the following scenarios:
   a) migration to space that has been driven
   b) migration to space that has been forced
   c) migration to space that is voluntary.

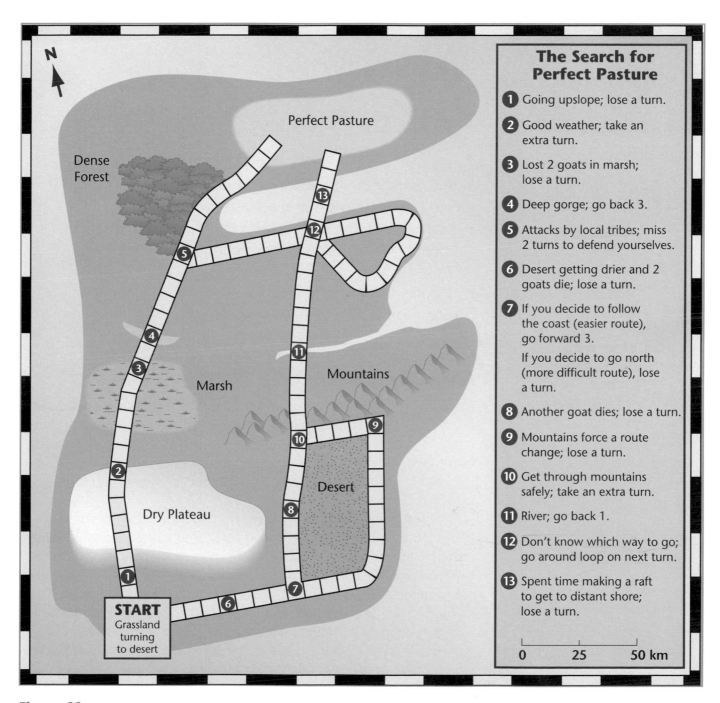

The Search for Perfect Pasture

1. Going upslope; lose a turn.
2. Good weather; take an extra turn.
3. Lost 2 goats in marsh; lose a turn.
4. Deep gorge; go back 3.
5. Attacks by local tribes; miss 2 turns to defend yourselves.
6. Desert getting drier and 2 goats die; lose a turn.
7. If you decide to follow the coast (easier route), go forward 3.
   If you decide to go north (more difficult route), lose a turn.
8. Another goat dies; lose a turn.
9. Mountains force a route change; lose a turn.
10. Get through mountains safely; take an extra turn.
11. River; go back 1.
12. Don't know which way to go; go around loop on next turn.
13. Spent time making a raft to get to distant shore; lose a turn.

0   25   50 km

Figure 11
Take turns rolling dice to move along the trail. The first to reach Perfect Pasture wins!

Chapter 3

Migration Patterns

In this chapter we focus on migration patterns in the world today. The information and activities will help you

▸ identify and describe patterns and trends in migration
▸ identify factors that influence people to move away from a place
▸ identify factors that influence people to move to another place.

Migration Around the World

In the world today, over 120 million people are living outside the country of their birth or citizenship. These people make up about 2 per cent of the world's population, and the number is growing every year. However, migrant people are not evenly spread around the world, rather the majority of them settle in only a handful of countries.

We learned in the last chapter that the movements these people have made can be grouped into such categories as "driven," "forced" and "voluntary." In this chapter, we will see patterns in the locations and directions of these movements. The four patterns we want to understand are shown in four colours in Figures 1 and 2 (bright red, light red, green and yellow). The bright red pattern —people who have migrated to another place within their own country because of war—is the most difficult to track. Some experts say that the number of such migrants is as low as 4 million. Others say it is as high as 50 million.

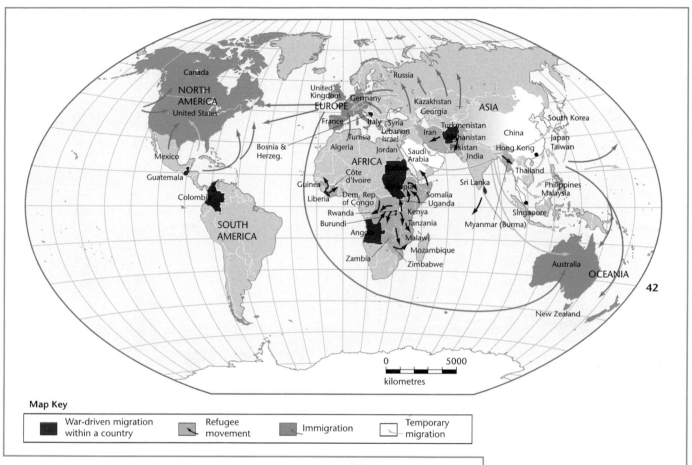

Figure 1
Migration Patterns Around the World

Map Key

■	War-driven migration within a country		Refugee movement		Immigration	□	Temporary migration

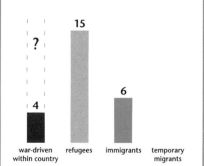

Figure 2
Estimates of the Number of Migrants in Each Migration Category (in Millions)

Discover ✺ *With Graphs*

1. Look at Figure 2. Why do you think experts have difficulty agreeing on the number of war-driven migrants within their own countries today?

2. Figure 3 on page 210 shows which regions of the world have more migrants entering than leaving, and which have more migrants leaving than entering.

a) Divide the six regions into "sending regions" and "receiving regions."
b) Compare Figure 3 with Figure 1. For each sending region, list one country migrants go to. For each receiving region, list one country migrants come from.
3. Figure 3 shows that Africa does not send out a very large number of migrants to other regions. But at least one-quarter of world migration takes place in Africa. How does Figure 1 help explain this?

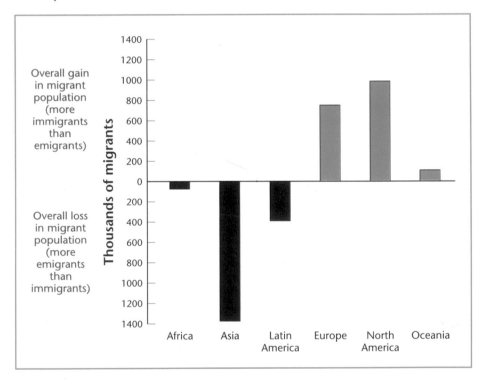

Figure 3
Migrant Gains and Losses by Region, 1990–1995

Migration Within a Country

In addition to the many millions of people who have migrated to *another* country, an estimated 4 to 50 million people have moved to another part of *their own* country. Figure 1 shows eight countries in which this movement has been especially great in the 1990s: Guatemala, Colombia, Bosnia/Herzegovina, Liberia, Angola, Sudan, Afghanistan, and Sri Lanka.

In most of these countries, violent conflict or actual **civil war** between opposing groups has been going on for decades. For example, in Colombia there have been 30 years of fighting between leftist guerrillas and right-wing paramilitary groups. In Africa and Asia this fighting usually leads to people moving both to safer parts of their own country and fleeing to other nations as refugees. For example, the many people fleeing their homes because of civil war in Liberia in the 1990s included at least 750 000 people who left the country.

Sudan—the largest country in Africa—has the largest number of people who have migrated within a country. This internal migration has been a pattern for a long time. For example, in the 19th century, 2 million slaves were captured in southern Sudan and taken to the North. The following case study presents migration patterns within the country today.

civil war—armed conflict between people within a nation. It may be started by a group that wants to take power from the current rulers or by a group that wants another group to leave the country.

Case Study
Staying and Leaving in Sudan

The history of Sudan has involved a variety of ethnic and religious groups. One large ethnic/religious group is made up of Arabic-speaking Muslims, who emigrated southward into Sudan from Egypt centuries ago. These people make up the most powerful group in northern Sudan. The country's government is in their hands. The southern population is made up of 600 ethnic groups who speak 400 languages and mostly follow traditional African or Christian religious practices.

The people of the South feel they have gained little from being part of Sudan. While the country overall is poor, services and living conditions are the worst in the South. From 1955 to 1972, southerners fought a civil war against the forces of the North. They wanted to gain independence as their own country. Another war beginning in 1983, aimed at getting fair treatment for all Sudanese people, continues today. More than 2 million southerners have fled the war and now live in the North.

Imagine the **push factors** that would affect you if you lived in southern Sudan. Let's say you are a member of a subsistence farming family. For centuries, your people have grown sorghum crops, herded cattle and lived in villages of thatched huts. The civil war between northern

push factors—the social, political, economic, and environmental forces that drive people away from one location to search for another one.

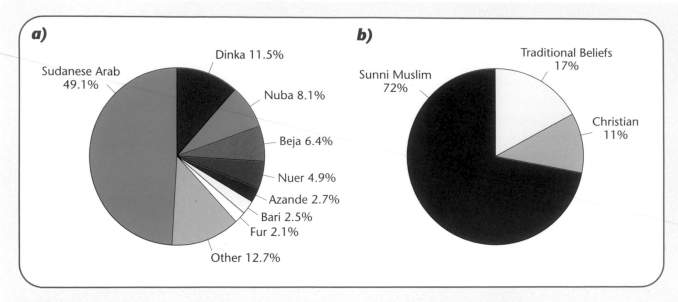

Figure 4
Ethnic (a) and Religious (b) Groups in Sudan

government forces and rebels in the South has been going on all your life. When you were a baby, many of your relatives died because of a great famine. Rebel bands have raided your village more times than you can count. There have been many food shortages because your animals were stolen and crops were destroyed. Last year there was no clean water, and many died from cholera. The nearest schools, hospitals, and roads are closed. Food aid from the United Nations and other organizations has helped to feed your people from time to time. Recently, your village has been a storage centre for grain from the UN World Food Program. You expect thousands of hungry people from other villages to arrive soon. You finally decided to move north when bombs from government cargo planes were dropped on your village.

Figure 5
The civil war is an ongoing battle in southern Sudan.

Migrant Life in the North

About half of those who have migrated to the North live around Khartoum, the country's capital. They work at low-paying jobs such as cooking and cleaning in other people's homes. They may live in squatter settlements in shelters made out of bits of tin, plastic and cardboard. Many sleep in the streets. In camps in the desert around Khartoum, there are kilometre-long rows of flat-roofed, mud-brick houses without a tree or bush in sight. Many do not mind these conditions if they can get work. But Arab residents see the migrants as a threat to their own employment and wage rates, so they want them to leave.

Figure 6
Southern Refugees Camped in Northern Sudan

Discover For Yourself

1. In small groups, imagine you are members of a subsistence farming family in southern Sudan.
   a) List four push factors that would make you want to escape from your village.
   b) List two **pull factors** involved in your decision to move.
   c) Review the barriers to migration on pages 202 to 204. Which barriers would be involved in your situation? Describe how they apply.
2. In small groups, discuss how Sudan's southern population is different from its northern population. Make point-form notes, describing at least two differences in detail. Discuss how these differences make it likely that the war will continue for a long time.
3. Look at Figure 6 on page 188. Which diagram in the figure applies to Sudan? How does this diagram relate to the civil war and to your answer to Question 2?

pull factors—the social, political, economic, and environmental attractions of new areas that draw people away from their locations.

LANGUAGE
LINK

Refugee Migration

Millions of refugees were driven or forced to migrate to other countries in the 1990s. Most of them moved to the closest safe area in a neighbouring country, as shown in Figure 1 on page 209. Both the source countries of refugees and the *asylum* countries to which refugees go are usually in developing regions.

Figure 7 shows a recent example of refugee movement. It involves ethnic Albanians living in a southern republic of Yugoslavia called Kosovo. In 1990, ethnic Albanians in Kosovo made up 90 per cent of the region's population. Many of them wanted to form their own nation. But the Yugoslav government was determined to maintain control of Kosovo. Fighting in the region started in 1991. Throughout the 1990s, Yugoslav government forces persecuted the ethnic Albanians, killing them or forcing them to flee to neighbouring countries. In 1999, NATO (the North Atlantic Treaty Organization, made up of Canada, the United States and many European countries) got involved. NATO military planes attacked key locations in Kosovo and elsewhere in Yugoslavia. They wanted to cripple the Yugoslav forces in Kosovo and also make life uncomfortable for people in Yugoslavia. The many push factors in Kosovo sent 700 000 refugees into Macedonia and Albania. Some were smuggled by boat to Italy. As tent camps near the Yugoslav border overflowed, many refugees were moved to other countries in Europe, as well as to North America.

Figure 7
Refugees may settle in a neighbouring region or be resettled in a more distant country with the help of international organizations. Once their home regions are safe enough, many are *repatriated* (returned home).

Immigration and Temporary Migration

Many of the world's immigration and temporary migration patterns can be explained by looking at population patterns. The work force in the developing world has reached a size of over 2 billion people. Every year in the developing world another 80 million people reach working age and join this enormous group. In Mexico, Turkey, and the Philippines, up to 1 million new jobs must be created every year to employ the young adults who enter the working world.

The chance to work is a very powerful pull factor for these millions of people. Industrialized countries exert the strongest pull on them. This explains why seven of the world's wealthiest countries (Germany, France, United Kingdom, United States, Italy, Japan, and Canada) have about one-third of the world's migrant population. Within the world's three "receiving regions" (see Figure 3 on page 210), several patterns can be seen:

▶ Europe: population gains from immigration occur in *northern* and *western* Europe. *Southern* and *eastern* Europe lose population from emigration.

▶ North America: Canada has one of the world's highest immigration rates, with over 40 per cent of the population growth coming from immigration.

▶ Oceania: population gains from immigration occur in *Australia* and *New Zealand* only. As with Canada, over 40 per cent of Australia's population growth comes from immigration.

Many people living in these receiving regions want to reduce immigration. See what the consequences of this attitude are in the following case study.

Case Study — *Managing Migration in Germany*

Germany, one of today's powerful and wealthy nations, is a magnet for immigrants and refugees. Following World War II, large parts of the country had been destroyed by bombs and other weapons of war. There was a shortage of workers to help rebuild Germany's industries, roads, and buildings. Foreign workers—mainly from Turkey and Yugoslavia—were invited to come help. This was Germany's first modern immigration wave. The second occurred in 1989 and 1990, when East and West Germany were reunited and the USSR was crumbling. People were able to travel more freely into Germany than ever before. As of 1997, yet more immigrants from other countries of the European Union have moved to Germany.

Like many other countries in Europe, Germany needs immigrants. Without them, its population would decrease at an alarming rate. One reason for this is that people are choosing to be childless or to have only one or two children. (On average, each German woman only has 1.3 children. To keep a population steady, each woman, on average, should bear 2.1 children.) A second reason is that the population of Germany, like that of Canada, consists of more and more older people. These people receive state pensions. Foreign workers are needed to help pay taxes that will contribute to these pensions and support the country's health system.

Some German people show prejudice and **discrimination** toward the country's immigrants. This behaviour exists despite laws to discourage it. Figure 8 shows the number of crimes in Germany that have sprung from prejudice against foreigners. In many cases, these crimes consisted of setting fire to foreigners' homes.

Usually the people who commit these crimes are afraid that the migrants who have moved into their neighbourhoods will take their jobs. But statistics show that this does not happen. In fact, the money that immigrants spend actually helps the economy to improve. As people begin to understand this, they may come to accept and even enjoy the cultures of their neighbours.

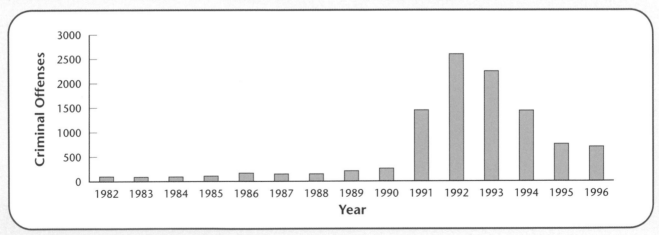

Figure 8
Criminal Offenses Against Foreigners in Germany Due to Prejudice, 1982 to 1996

discrimination—treating a group or individual unfairly based on their background.

 **With Maps and Graphs**

1. Figure 9 shows the numbers of people who entered Germany in 1996.
   a) On an outline political map of Europe and western Asia, name and lightly shade in (using one colour) each of the countries listed in Figure 9.
   b) Choose three colours different from the colour used in (a). One colour will represent foreigners, a second colour will represent refugees and a third colour will represent ethnic Germans. Use these colours to draw arrows that show the movement of people from their country of origin into Germany. Use wider arrows for heavier flows of people and narrower ones for lighter flows of people.
   c) Write the actual numbers of people in a small circle where each arrow begins in the country of origin.
   d) Finish your map with a suitable title, legend, scale and north sign.

Country of Origin	Foreigners	Refugees	Ethnic Germans*
Poland	77.4		1.2
Turkey	73.2	23.8	
Italy	45.8		
Yugoslavia	42.9	18.1	
Portugal	32.0		
Russia	31.9		172.2
Romania			4.3
Iraq		10.8	
Afghanistan		5.7	

*People who are of German descent

Figure 9
Inflow of Foreigners, Refugees, and Ethnic Germans into Germany, 1996 (in Thousands)

2. Compare the graphs in Figure 8 on page 216 and Figure 10 below. In small groups, discuss any patterns you see. What do you think might explain these patterns?

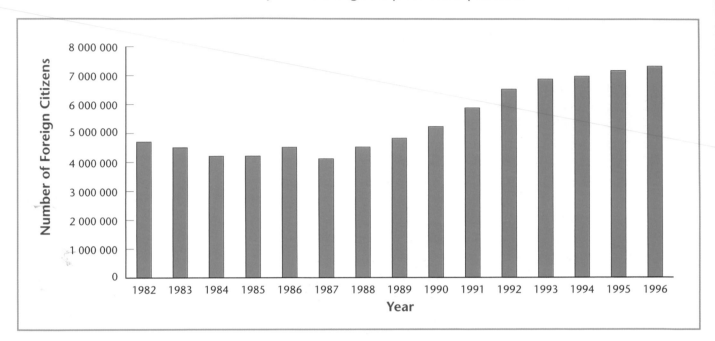

Figure 10
Number of Foreign Citizens in Germany, 1982 to 1996

Summary

In this chapter you have discovered patterns involving internal migrants, refugees, immigrants, and temporary migrants. You have learned about the push factors that cause internal migration and refugee movements. You have also seen the advantages and disadvantages of immigrating to another country for work.

Reviewing Your Discoveries

1. Explain the meaning of "push factor." Give two examples.
2. Explain the meaning of "pull factor." Give two examples.
3. How is internal migration in Sudan similar to refugee movement from Kosovo?
4. Give three reasons why Germany has large numbers of immigrants.

Using Your Discoveries

1. In small groups, research a situation that is currently causing many people to leave their homes. The situation could involve permanent or temporary migration. You may wish to use Figure 1 on page 209 as a guide for finding a situation. Newspapers, magazines and the Internet will also be helpful. Produce a visual display that includes the following:

    a) a map to show the area where the people are moving away from and where they are moving to
    b) statistics on the number of people involved
    c) a description of the push and pull factors involved
    d) a mock written "interview" (about 3 paragraphs long) with a small group of migrants right after arriving in their new home. The "interview" should include their recent experiences in their own words.

Figure 11
During the summer of 1999, two boatloads of Chinese refugees arrived on Canada's west coast. The newcomers shown in the photograph were smuggled to the Queen Charlotte Islands in a fishing trawler and then taken to Port Hardy by a Canadian government vessel. Both boatloads of refugees were from Fujian province in southeastern China. In 1998, nearly 100 000 Fujians (at least one family member from nearly every household in the province) were smuggled overseas.

WEB LINK

To find out more about human migration, look up
http://www.nationalgeographic.com/features/2000/population/migration

Chapter 4

Coming to Canada

Key terms

open-door immigration
bilingual
official languages

open-door immigration—
immigration that is free or
unrestricted.

In this chapter we focus on immigration patterns in Canada. The information and activities will help you

▶ describe patterns and trends in immigration to Canada
▶ use graphs and maps to identify the sources and settlement patterns of immigrants to Canada
▶ show an understanding of the effects that migration has had on the development of Canada.

Immigration Patterns

Of all the cultural groups in Canada, Aboriginal peoples have been here the longest. They have seen their lands and their lives transformed by those who came later.

Immigrants from Europe first arrived in Canada in the 16th century. They were mainly French. In 1759, British forces defeated the French in a war in Quebec. From that time onward, British settlers arrived in large numbers. Significant numbers of European immigrants started to arrive in the mid 1800s. Figure 1 shows the major waves of immigration from 1870 to 1989.

Before Confederation in 1867, Canada had an **open-door immigration** policy. Anybody who wanted to, and who could afford the journey, could come to Canada. At that time, about 90 per cent of Canada's population was either British (60 per cent) or French (30 per cent).

During the next century, the government developed a more restrictive immigration policy. It favoured people from the United

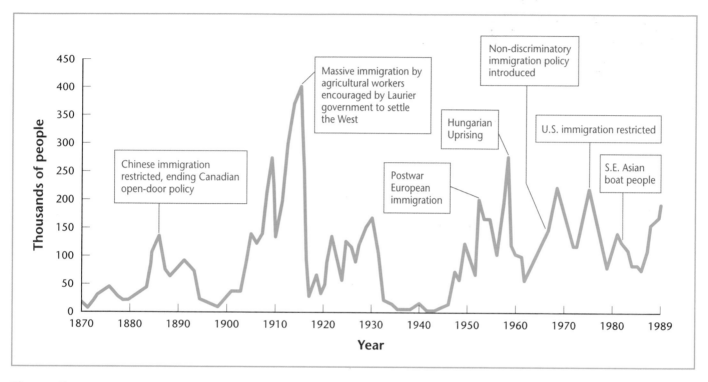

Figure 1
In the 20th century, immigration levels in Canada were lowest during the worldwide Great Depression (1930s) and World War II (1940s).

Kingdom, the United States, and Western Europe. Canada had few immigration offices in Asia, Africa and the West Indies, so it was difficult for people in these areas to apply to become immigrants. Laws regulated the number of American Blacks, Indians, and Japanese immigrants allowed into the country. Canada even turned back Jews fleeing Nazi persecution during World War II. Other groups not allowed into the country included communists, labour organizers, people with infectious diseases, and people with physical or mental handicaps.

The 1885 Chinese Immigration Act included a "head tax" (a fee for every Chinese newcomer) to discourage immigration from China. In 1903, the $500 head tax was equivalent to $10 000 in today's money. This meant that usually only one member of a family, often the father, could afford to come to Canada. It was a lonely life for the men in their new country. They worked hard to send money home to support their families. In 1923, the Canadian Parliament passed an even more restrictive law that virtually prevented *any* Chinese immigrants from entering the country. This discriminatory law was *repealed* (reversed) in 1947.

In the late 1950s there were few requirements which had to be met by British, French and American applicants, as a result many immigrants entering Canada were unskilled workers from Europe. In 1962 new regulations were introduced that made education and occupational skills the number one criteria for admitting new immigrants, and the special provisions for British, French, and American immigrants were dropped.

In 1967, a "non-discriminatory" immigration policy was introduced. In this policy, a "points system" was created to evaluate immigrants from all countries by the same standards.

In 1971, the government declared that Canada was "multicultural within a **bilingual** framework."

In 1978, Canada grouped immigrants into three types: refugees, family, and independent. The points system from 1967 is still used to help choose immigrants in the independent class. Since 1992, the Canadian government has favoured independent immigrants who (a) have business skills or money to invest in Canada; and (b) will adapt well to Canadian society.

Each year, the government sets a limit on how many people will be allowed into the country. Once the limit is reached, even those who have the 70 points needed to qualify as independent immigrants may not be accepted. For example, in 1990, *1.5 million* independent families applied to become immigrants. In that year, the government decided to accept a *total* of 250 000 immigrants, with a limit of *20 000* independent families. (The other limits were: 21 000 sponsored families, 10 000 refugee families and 4000 entrepreneur families.)

Figure 2 shows how Canada's changing immigration policies have resulted in varying numbers of immigrants from different parts of the world. Figure 3 shows the source areas of immigrants now in Canada.

bilingual—involving two languages. In Canada, this term refers to French and English.

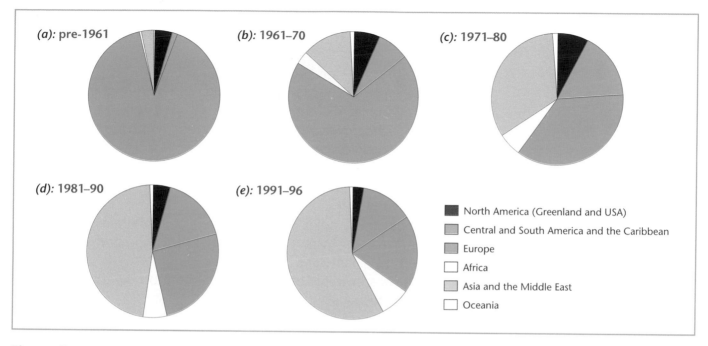

Figure 2
Immigrants to Canada by Region of Origin

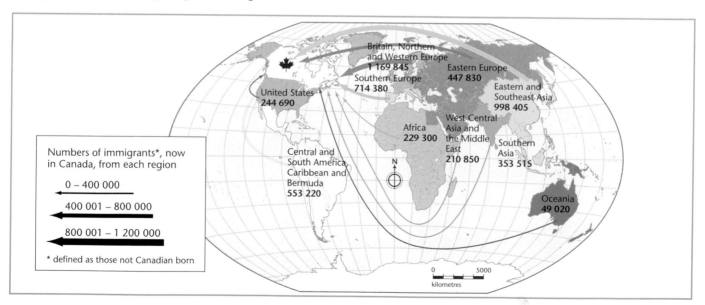

Figure 3
Number and Origin of Immigrants Living in Canada, 1999

Discover  With Graphs

1. Look at Figure 1 on page 221.
   a) In which year was the highest number of immigrants received in Canada? How many were there?
   b) Between which years was there a lengthy period when Canada had very few immigrants?
   c) Suggest two reasons for the lack of immigration during
      – the Great Depression
      – World War II
   d) List two examples of high numbers of immigrants due to *push factors*. Include the date of each. Suggest why people left their homelands in each case.
   e) List one example of a *pull factor* written on the graph. Explain why this is called a pull factor.
2. List and briefly explain three different examples of Canada's restrictive immigration policies from 1867 to 1967. What kinds of barriers to immigration did these policies present? Explain your answer.
3. Use Figure 4 to make your own line graph of Canadian immigration from 1990 onward. Get more recent data to add to these figures if possible.
   a) How do the number of immigrants compare with those in previous years shown in Figure 1?
   b) Compare the variation in the number of immigrants in your graph to the variation in Figure 1. Suggest a reason for the difference.

Turn to page 211 and 213 to review push and pull factors. Turn to page 202 to review barriers to migration.

MATH LINK

Year	Immigrants
1990	213 334
1991	232 020
1992	253 345
1993	255 935
1994	223 912
1995	212 463
1996	226 074
1997	216 044

Source: Statistics Canada

Figure 4
Number of Immigrants to Canada, 1990–1997

4. Look at Figures 2 and 3 on page 223.
   a) Which group in the immigrant population has decreased very significantly?
   b) Where do the majority of today's Canadian immigrants come from?
   c) Around what time did this group become the largest?
   d) Use an atlas or other source to find the GNP per person and population density in Canada, the United States, Australia, and New Zealand. How do they compare? How does this help to explain why there are fewer immigrants from the United States and Oceania?

Figure 5
In 1993, the nearly 256 000 immigrants to Canada formed one of the highest numbers of newcomers in Canadian history. The record of 400 000 occurred in 1913, when this photograph was taken.

Where Immigrants Go

Most of the immigrants who come to Canada today live in large cities. Why is this? One reason is that more jobs are available in cities. A second reason is that the population of large cities usually already includes members of the same cultural group as new immigrants. It is important to be with people to whom you can talk and relate. It is especially important to newcomers who have not learned to speak English or French.

Toronto attracts more immigrants than any other city in Canada. Over 30 per cent of all recent immigrants to Canada (1991–96) live in Toronto. Forty-two per cent of the people in Toronto were born in other countries. There are at least 1000 immigrants from each of 100 countries living in the city. As for the rest of Canada, Figure 6 shows which provinces and territories immigrants said they were going to on first arriving in Canada in 1997.

Turn to page 86 to review GNP per person.

YK	NWT	BC	AB	SK	MN	ON	QU	NB	NS	PEI	NF	Total
100	86	47 459	12 919	1 742	3 804	118 060	27 672	663	2 891	151	437	216 044

Figure 6
Immigration by Region of Intended Destination, 1997

Discover With Graphs and Maps

MATH
LINK

1. Draw a bar graph to represent the data in Figure 6 on page 225.
   a) What percentage of the 1997 immigrants came to Ontario?
   b) Rank the top four provinces with respect to the destination of immigrants.
   c) What do you believe makes these provinces more attractive than the other provinces and territories?
2. Figure 7 shows the settlement patterns of five recent immigrant groups in Toronto. Figure 7 (f) is a locator map for Toronto areas. Divide into small groups to analyze the maps and answer these questions.
   a) For each group, describe the locations where the recent immigrants are concentrated. Use Figure 7 (f) to help you.
   b) Give three reasons why people in a specific group would be attracted to one part of the city over another.

Figure 7
Settlement Patterns of Selected Immigrant Groups in Toronto, 1991–96 (1 dot = 20 people)

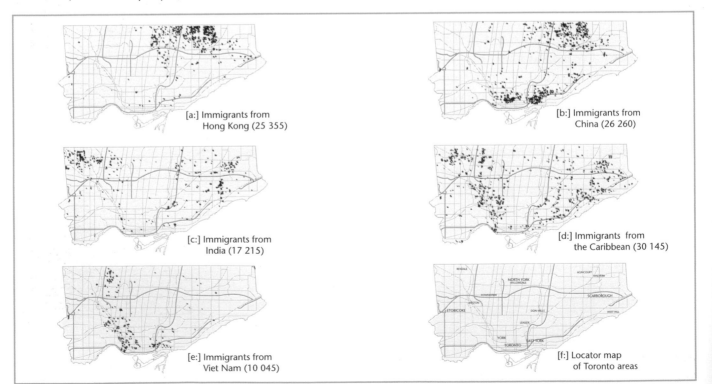

[a:] Immigrants from Hong Kong (25 355)

[b:] Immigrants from China (26 260)

[c:] Immigrants from India (17 215)

[d:] Immigrants from the Caribbean (30 145)

[e:] Immigrants from Viet Nam (10 045)

[f:] Locator map of Toronto areas

Contributions to Canada

Immigrants and their descendants have had a great effect on Canada. Here is an overview of immigrants' contributions to Canada through history.

Starting in the middle 1500s, Europeans were attracted to Canada by the abundance of fish off the country's east coast. At first they came ashore only to salt and dry the fish to preserve it for the homeward journey. Many eventually decided to settle in small fishing villages along the coast. They brought with them their knowledge of fishing technology, and a fishing industry has continued since that time.

As French and British settlers arrived in greater numbers, they began the process of developing the Canadian identity. These two nations gave Canada its **official languages** of French and English. They were also responsible for the way our government is organized.

Immigrant farmers in Canada preferred to make farms on the kind of land they farmed back home. In the late 1800s and early 1900s, land in the Prairie provinces was offered free to those who agreed to farm it. This was a strong pull factor, especially for Ukrainian immigrants. Many of the Ukrainian immigrants were expert wheat growers and were attracted to the flat land and rich grassland soils so much like those of their homeland. Dutch farmers were experienced with making wet marshland soils usable. Immigrant farmers drained Ontario's Holland Marsh and created a very successful vegetable growing area north of Toronto. This area still flourishes today, and many of the farms are now operated by the descendants of the first Dutch farmers.

Much industrial "know-how" was brought to Canada by immigrants. Many of them knew how to build railways, canal locks, and mines, as well as flour, saw, and textile mills. Some of these structures had to be changed to operate in Canadian conditions. For example, some waterwheels were put within mill walls to prevent them from being damaged by snow and ice. Traditional European ploughs were altered so they could be manoeuvred around tree stumps and rocks. Ships were designed to travel on the Great Lakes and on shallow rivers.

official languages—the languages in which the federal government conducts its business. Some provincial governments and many businesses use both official languages in their written material.

Figure 8
Holland Marsh supplies fresh vegetables to southern Ontario.

Figure 9
A Laker in Welland Canal, Ontario

Canada's first railways were built by immigrants, particularly those from China. Immigrants mined the iron ore needed and made the steel rails. The railway enabled people to settle in various places, develop the resources there and ship out their products.

Today, the government selects the immigrants Canada needs to help the country develop further. On average, today's immigrants are better educated than their Canadian counterparts. For example, of the immigrants who arrived between 1991 and 1996, 34 per cent of those aged 25 to 44 had completed university. In comparison, only 19 per cent of the Canadian-born population in the same age group were university graduates. Almost 90 per cent of these immigrants had studied in the fields of science and technology.

The knowledge, training, and experience immigrants bring make them a vital part of our work force. For example, in Toronto, immigrants make up

▸ 58 per cent of chemists
▸ 33 per cent of school teachers
▸ 63 per cent of auto workers
▸ 45 per cent of physicians and surgeons
▸ 32 per cent of sales staff
▸ 58 per cent of those involved in food and hospitality services
▸ 37 per cent of office workers

And these are just a few examples.

Working immigrants pay taxes that help to run services such as education, health care, and pensions. They spend money on housing, food, and consumer goods, which helps other industries to survive.

When immigrants first arrive, their unemployment rate is usually higher and their incomes are usually lower than those of the general population. But once they have had a chance to find work, their unemployment rate is lower than average, while their incomes are higher than average. They also tend to use welfare much less often than Canadian-born workers. This situation may be changing, as statistics show that immigrants had less success finding jobs in 1996 than they did in 1986.

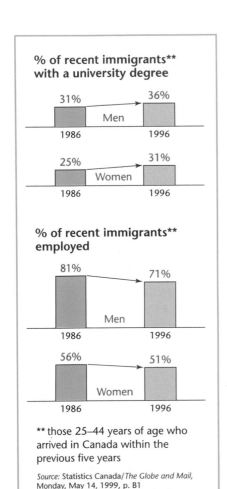

% of recent immigrants** with a university degree

31% — Men — 36%
1986 — 1996

25% — Women — 31%
1986 — 1996

% of recent immigrants** employed

81% — Men — 71%
1986 — 1996

56% — Women — 51%
1986 — 1996

** those 25–44 years of age who arrived in Canada within the previous five years

Source: Statistics Canada/*The Globe and Mail,*
Monday, May 14, 1999, p. B1
Credit: Carrie Cockburn/*The Globe and Mail*

Figure 10
Immigrant education levels have risen, but employment levels have sunk.

Case Study

An Immigrant Family in Toronto

Mr. Rajiv Kashi moved to Toronto from a small town in Punjab, India in the late 1960s. He was a post-graduate student in business administration. In 1974, his younger brother joined him and attended university for four years. The two brothers started a business selling electric light fixtures.

The brothers' parents, who were still living in India, chose wives for the brothers. The brothers met and married these women on a trip to India and returned to Canada with their new brides.

In the early 1980s, the two brothers sponsored their parents and a younger brother as immigrants to Canada. Their father, Mohan, helped them to buy a six bedroom house in northern Toronto. Now living in the house are three generations of the family:

▸ Mohan and his wife Shivani
▸ Rajiv, his wife, his two brothers and his brother's wife
▸ the four children born to Rajiv's and his brother's families

The family is economically successful and has adapted well to Canada. They keep some of their traditional culture. For example, they are vegetarians and follow the Hindu religion. Shivani and her two daughers-in-law observe the annual fast (called *karva chauth*) for the well-being of their husbands.

Once the children went to school full-time, their mothers went to work outside the home as a secretary and a sales clerk. They wear the same style of clothing as other people at their workplaces. Only on special occasions at home do they wear the traditional clothing of a *saree* or *salwar-kameez*. The women and men speak Punjabi among themselves, but they do not expect the children to learn the language. This makes Shivani, the children's grandmother, feel left out because she doesn't understand English well. She also wishes that her daughters-in-law would wear traditional Indian clothing all the time, as she does.

The children are all teenagers now. Their grandparents are bothered about the fact that they refuse to wear Indian clothes anywhere and eat hamburgers when they go out with their friends (eating beef goes against Hindu traditions). The middle-generation parents are worried about their children losing their culture. They would much prefer them to follow the old traditions. But they realize it will be difficult to force their point of view on their children as the children get older.

Figure 11
One of several Indian Communities in Toronto

Discover  For Yourself

1. In small groups, discuss these questions about the case study.
    a) Identify three things that are changing from one generation to the next in the Kashi family. Explain what is happening in each case.
    b) Explain why you think the changes are taking place.
    c) What do you think Mohan's and Shivani's *great*-grandchildren will keep of their Punjabi heritage? Do you consider this to be good or bad? Explain your answer.

Summary

In this chapter you have learned that Canada's population is composed of Aboriginal peoples, immigrants, and descendants of immigrants. You have discovered which groups have come to Canada at different times in its history. You have also seen that immigrants contribute to Canada although they also face challenges finding work and maintaining their cultures.

Reviewing Your Discoveries

1. From which continent do most of Canada's immigrants come today? How is this different from the sources of Canadian immigrants before 1961?
2. Explain three ways in which immigrants
    a) helped Canada to develop in the past.
    b) play an important part in society today.

Using Your Discoveries

1. Look at the Restaurant Guide in Figure 13.
    a) List the countries or parts of the world whose food is represented.
    b) On a world map, label and colour the countries or areas. Finish your map with a suitable title.
    c) Explain how the presence of immigrants from these countries and areas contributes to towns and cities in Ontario.

Figure 12
Adrienne Clarkson was born in Hong Kong, of Chinese parents. She and her parents came to Canada as refugees in 1942. In 1999 she became Canada's 26th Governor General.

2. Imagine the Yellow Pages for an average Ontario city in 1950. List the kinds of restaurants that you would have expected to find there in that year. Use Figure 2 on page 223 to help you.

Figure 13
This is a made-up restaurant guide for an average Ontario city.

RESTAURANT GUIDE

RESTAURANTS ARE LISTED ALPHABETICALLY BY NATIONALITY

AUSTRIAN

Lucas's Place
94 John Street 555-1873

Bruk and Company
16 Freeman Drive 555-9264

BRITISH

Fox and Hound Pub
50 Sterling Street 555-8362

Fawlty Towers
29 Torquay Road 555-6721

Heather and Hide Pub
52 Scotia Place Road 555-6734

The Printer's Devil
29 East Harding Street 555-6400

The Queen's Place
777 Osgood Avenue 555-4343

The White Hart
1610 Runnymede Avenue 555-8201

CANADIAN

Mountie's Diner
106 Riverside Drive 555-2281

Maple Creek Inn
42 Elm Street 555-1537

Moosehead Café
42 Church Street 555-1537

CARIBBEAN

Island Restaurant
1943 Fort Street 555-7632

Beaches
15 Birch Avenue 555-0012

CHINESE

Golden Dragon Restaurant
10 York Street 555-1272

Jade Palace Restaurant
1900 Front Street 555-4466

Lucky Garden Restaurant
45 Eighth Street 555-4637

MING GARDEN BUFFET HOUSE
Chinese Buffet 7 days a week
Over 50 items to choose from
Call for reservations **555-0043**
67 Sixty-first Avenue

Oriental Garden
160 – 14th Avenue 555-2200

Peking Take-out
240 AlbertStreet 555-4466

Phoenix restaurant
190 Front Street 555-1100

Wong's Take-out
59 Peter Street 555-6677

EGYPTIAN

Ra Restaurant
40 Greenlane Street 555-9021

The Kings' Café
898 Cross Street 555-3342

FRENCH

C'est Bon Café
54 Ramsay Avenue 555-8237

Yves' Bistro
89 Common Court Street 555-7321

GERMAN

Dana's Bakery and Café
13 Auld Road 555-9018

BRATWURST CLUB RESTAURANT
Authentic German Cuisine
Huge selection of imported beer
456 Peter Street **555-0291**

ITALIAN

The Great House of Pasta
563 First Avenue 555-4635

Pizza Palace
77 Eighth Street 555-3122

The Ripe Tomato
61 Peter Street 555-3476

Dino's Diner and Take-out
1002 Ramsey Avenue 555-0980

JAPANESE

Sushi Garden
899 Elm Street 555-9090

House of Tempura
421 John Street 555-7650

Sake Restaurant
12 First Avenue 555-8122

MEXICAN

Sombrero
67 Peter Street 555-1414

Tequila Club Café
190 Common Court Street 555-3488

Chapter 5

Going the Distance

Key terms

modes of transportation

tourist

travel hub

modes of transportation—
the specific means by which
people move from one place to
another (for example, on foot,
by bicycle, by car).

In this chapter we focus on patterns in people's regular travel movements. The information and activities will help you

▶ identify patterns in the use of cars and other modes of transportation
▶ describe the challenges of commuting in a big city
▶ use a decision-making model to choose an ideal place to visit.

Modes of Transportation

Think about how you got to school this morning. How many **modes of transportation** did you use?

When Muhammad goes to school, he uses a wheelchair to get to the end of his driveway. A bus picks him up and stops along the way to pick up other passengers. Muhammad then travels by wheelchair from the bus to the school building. These three stages make up a "journey chain." This journey chain has three stages or "links": wheelchair, bus, wheelchair. At the end of each link, a change in the means of transportation occurs. This point is called a "junction."

Most of the travel time in the chain is spent on the bus, although getting into and off the bus also takes a long time. Since the bus picks up three other students in wheelchairs, the overall journey is very slow for the fairly short distance travelled. Figure 1 shows Muhammad's route to school. Figure 2 shows the time it takes to travel different parts of the journey.

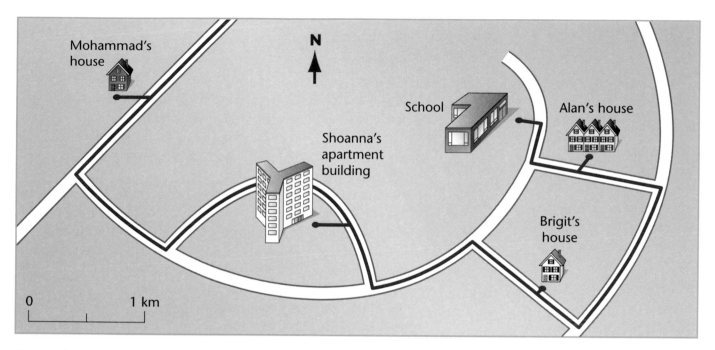

Figure 1
Muhammad's route is a journey chain of three stages: wheelchair, bus, wheelchair.

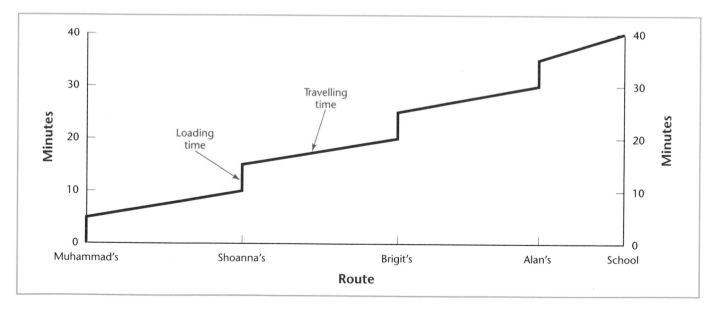

Figure 2
This graph shows how time is spent in Muhammad's journey to school.

Discover With Maps and Graphs

1. Look at Figures 1 and 2 on page 233.
   a) Measure the distance that Muhammad travels on the bus to school. State your answer in kilometres (km).
   b) How long does it take Muhammad to travel this distance? State your answer first in minutes and then in hours (h) (using decimals).
   c) How long would it take if one of the other students was absent?
   d) What is the average speed of the whole journey? State your answer in kilometres per hour (km/h). Use this formula to calculate average speed:
   Average Speed (km/h) = Distance (km) ÷ Time (h)
2. Think about your route to school.
   a) Draw a map showing your route to school and the different modes of transportation that you use. Use a differently coloured line to show each mode of transportation. Explain these colours in a legend. Include a line scale and north sign.
   b) Draw a line graph showing the time involved in various stages of your journey.
   c) Calculate the average speed you travel on your journey. Show your calculations.

MATH LINK

Figure 3
A drive through bank, Toronto

Global Patterns

People in developed nations such as Canada have the most choice when it comes to modes of transportation. In North America, the preferred method is clear: the distance people travel by car is *10* to *20 times* greater than the distance they travel by bus or train. The pattern is similar in Western Europe, except that buses and trains are used a little more there than here.

Many parts of our culture have been affected by our love for cars. Our highway landscapes, motels (*mo*tor ho*tels*), drive-through restaurants and banks, and high suburban populations can all be linked to a huge dependence on cars. Stock car racing is the fastest growing spectator sport in the United States.

This use of cars is possible because of our high level of economic development. Even people with below-average incomes often find ways to afford the use of a car. When we look at modes of transportation in poorer countries, we see a different story.

In China, the distance people travel by *bus* is three to ten times greater than the distance they travel by car. Statistics for the use of walking and biking are very difficult to get, but these "non-motorized" modes of transportation are used much more often in developing countries than in developed ones.

Figure 4
Bicycles and rickshaws (wheeled carts pulled by people) are very popular modes of transportation in Bangladesh, where most people cannot afford cars.

The world's poorest people travel very little. If they do need to go some distance, trains are used most often.

Commuting

Our choice of mode of transportation is affected by how long we want to spend travelling. For example, it takes an hour to travel from parts of downtown Toronto to a workplace in Don Mills by bus and subway. The same trip by car takes only half an hour. Researchers have found that people all over the world usually spend between one and one-and-a-half hours a day "commuting" —travelling between home and work. Their research included comparisons between African villagers, Japanese manufacturing workers, Canadian employees, and many others.

In cities, most of this commuting takes place in the early morning and late afternoon. On an average day in Toronto in 1996, about 1.2 million trips were made in the city between 6 and 9 a.m. Almost 5 million trips were made in total throughout the day. The modes of transportation were:

▸ Car 68% (54% by car *driver* + 14% by car *passenger*)
▸ Bus/subway 22%
▸ Commuter train 1%
▸ Bicycle/walking 8%

Case Study   *All Roads Lead to Toronto*

Traffic is heavy *to* and *from* Toronto as well as within the city itself. In an average 24-hour period in 1991, almost 2 million trips were made to work between municipalities in the Greater Toronto Area. Figure 5 shows the municipalities involved in these trips. Figure 6 shows the highway system commuters use.

The roads in Figure 6 are called "limited-access highways." This means that they can be entered or exited at only a limited number of places, using on- and off-ramps. Compared to normal roads, there are far fewer places where the traffic flow can be increased, decreased or interrupted in any way. This makes travel quicker and more efficient.

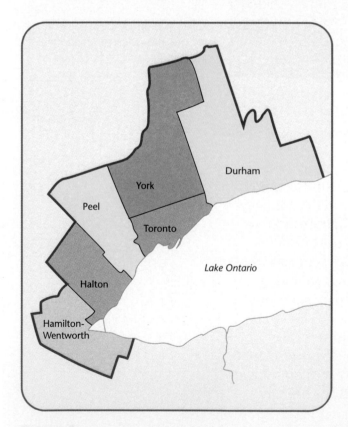

Figure 5
Municipalities Around Toronto

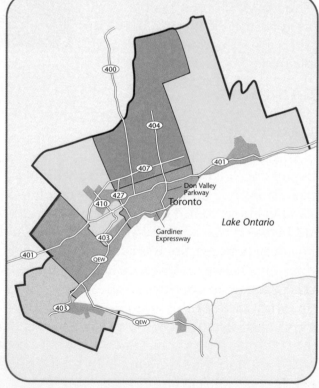

Figure 6
Major Highways in Toronto and Surrounding Municipalities

 **For Yourself**

1. Look at the data in Figure 7.
   a) How many people made a trip to work on one typical day in the Toronto area?
   b) Which destination were most workers going to?
   c) Rank the top four municipalities with the most workers going to the place you identified in (b).
   d) Look at the map in Figure 5. How does it help account for your answer to (c)?
   e) Do you think the municipality of Toronto should get funding from surrounding municipalities for road construction and repairs? Explain your answer.

| | To | | | | | | |
From	Toronto	Durham	York	Peel	Halton	Hamilton-Wentworth	Region Total
Toronto	789 500	11 500	67 500	62 000	5 500	1 000	**937 000**
Durham	48 500	92 000	8 500	2 000	500	0	**151 500**
York	94 000	3 500	85 000	9 500	500	500	**193 000**
Peel	106 500	1 500	11 500	173 000	7 500	1 500	**301 500**
Halton	24 000	0	2 000	24 500	60 500	9 500	**120 500**
Hamilton-Wentworth	6 000	0	1 000	5 000	20 500	121 000	**153 000**
Region Total	**1 068 500**	**109 000**	**175 000**	**275 500**	**94 500**	**133 500**	**1 856 000**

(Figures are rounded to the nearest 500. Therefore, some totals may not appear correct.)

Figure 7
Over an average 24-hour period in 1991, almost 2 million trips to work were made within and between municipalities in the Greater Toronto Area.

2. Form small groups. As a group, make a large copy of the map in Figure 5 on a piece of construction paper. Your map should be about 40 cm by 40 cm. Then follow these steps:

a) Choose six colours to represent the six municipalities listed in Figure 5. Lightly shade in each municipality on your map in its chosen colour.

b) Draw a circle in each municipality to represent travel within it (for example, travel **from** Peel **to** Peel). Make your circle larger or smaller depending on the number of trips taken. Follow guidelines such as those below:
   - Over 800 000 trips: radius of circle 5 cm
   - 100 000 to 800 000 trips: radius of circle 3.5 cm
   - 50 000 to 100 000 trips: radius of circle 1.5 cm

   Note: Fill in the circle with a darker shade of the colour chosen for the municipality.

c) Draw straight arrows between municipalities to represent trips from one municipality to another. Make your arrows wider or narrower depending on the number of trips taken. Follow guidelines such as those below:
   - Over 80 000 trips: arrow 5 mm wide
   - 60 000 to 80 000 trips: arrow 4 mm wide
   - 40 000 to 60 000 trips: arrow 3 mm wide
   - 20 000 to 40 000 trips: arrow 2 mm wide
   - 500 to 20 000 trips: arrow 1 mm wide

   Note: The colour of the arrow should match the colour of the municipality *from which* the trip was taken.

d) Finish your map with an appropriate title, legend and north sign.

3. Highway 407 in Figure 6 is a "toll" highway—people have to pay to use it. Since a limited number of people want to pay to drive on the highway, it has less traffic and offers a much speedier trip than other highways. But the 407 option has not been as popular as expected. Discuss these questions in small groups.

a) Do you think the drivers you know would use the 407 highway if it made their commute faster? Explain why or why not.

b) Working together, decide on one change to how the highway system in Figure 6 operates that would improve the commute of workers. Explain the reasons for your suggestion using Figure 5 or the map you made in Question 2.

c) What problems might arise as a result of your suggestion in (b)? How could these problems be solved?

Figure 8
These cameras are mounted over the on- and off-ramps of Highway 407. They pick up a signal from transponders or record the licence plates and times of cars entering and exiting the highway.

Richer, Further, and Quicker

As people earn more money, they usually travel greater distances. In 1960, the average North American earned $9600 and travelled 12 000 km in a year. By 1990, average income had doubled, and average distance travelled doubled right along with it. Since people travelling twice as far still don't want to spend much more than an hour getting there, they will start using faster modes of transportation.

In China, income tripled between 1960 and 1990. But people in 1990 were travelling *10 times* more kilometres a year than in 1960. In 1960, most Chinese people travelled only by walking or on bicycles. With more income, they could afford to use buses and trains, which are much faster modes of transportation. Also, cities grew a lot between 1960 and 1990, so many people lived further from their work and needed to commute further.

People in developing nations spend 3 to 5 per cent of their fairly low incomes on transportation. In more developed nations, where car ownership is common, people spend an average of 10 to 15 per cent of their income on transportation.

These patterns can help geographers predict the travel trends of the future. The first prediction is that the distances people travel will continue to increase as their incomes increase. Another prediction is that high-speed transport such as airplanes will be used for a greater proportion of travel in North America. At the same time, the slower modes of trains, buses, and cars will be used for a smaller proportion of travel. Worldwide, high-speed transport will probably be used for 40 per cent of the total kilometres the world's population will travel in a year. The third prediction is that most people will still spend between one and one-and-a-half hours travelling every day. In developing countries, buses will still be fast enough to meet people's daily travel needs. Among the super-rich, airplanes, often privately owned, will be used for even short journeys.

Transportation trends change slowly. This is because transportation *infrastructure*—roads, railways, airports, canals and so on—takes years to develop and costs a great deal of money. So it will be at least the late 21st century before geographers will deal with very different transportation patterns from those of today.

Discover With Graphs

1. Figure 9 shows changes in modes of transportation from 1960 to 2050.
   a) What is happening to the total number of passenger kilometres (pkm) travelled between 1960 and 2050?
   b) What is happening to the percentage of traffic made up of
      – buses?
      – trains?
      – automobiles?
      – planes?

c) Calculate the actual number of passenger kilometres (pkm) travelled by automobile (car) in 1960. Use the following formula:

number pkm by car = total number of pkm x percentage by car ÷ 100

5 500 000 000 000   x   54   ÷ 100

(see bottom of 1960 graph)

d) Calculate the actual number of passenger kilometres to be travelled by automobile in 2050. Use the same formula.

e) What is misleading about the appearance of the graph when it comes to showing the use of automobiles in 1960 and 2050?

f) Even in poorer countries it is true that as incomes increase so do the distances and the amount of money people spend on transportation. With that in mind, explain why the use of buses increases from 1960 to 2020.

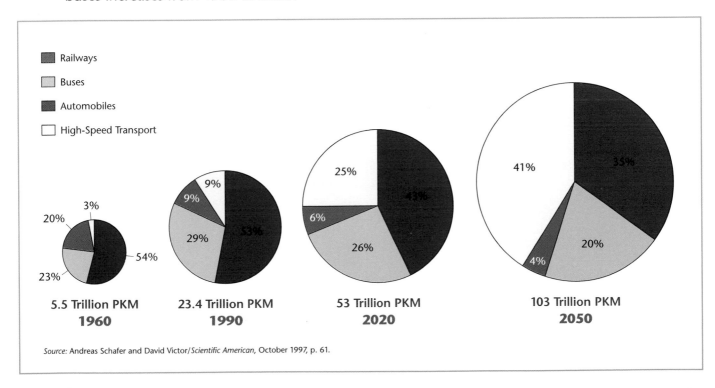

Railways
Buses
Automobiles
High-Speed Transport

5.5 Trillion PKM
1960

23.4 Trillion PKM
1990

53 Trillion PKM
2020

103 Trillion PKM
2050

Source: Andreas Schafer and David Victor/*Scientific American*, October 1997, p. 61.

Figure 9
In these graphs of worldwide travel patterns, "pkm" is short for "passenger kilometres," or the number of kilometres travelled by passengers.

Turn to page 260 to review making circle graphs.

2. Make your own circle graph of the kilometres you travel and the modes of transportation you use in an average week. Follow these steps.
    a) Fill in an organizer like the model shown in Figure 10.
    b) Write the kilometres travelled by each mode used as a fraction of the total. For example, from Figure 10 the list would be:
        – Bike   21/37
        (21 of the 37 kilometres travelled in a week are by bike—5 between Home and School and 16 between Home and Aunt's House)
        – Walking   6/37
        – Bus/Subway 10/37
    c) Calculate the number of degrees each mode will take up within a circle of 360°. Follow the model below, based on the information in (b).
        – Bike   21 x 360 ÷ 37 = 204°
        – Walking   6 x 360 ÷ 37 = 58°
        – Bus/subway 10 x 360 ÷ 37 = 97°
    d) Make your circle graph and write in all the labels needed. Use the circle graphs in Figure 9 as a model.

Places travelled:	Home ↔ School	Home ↔ Stores	Home ↔ Aunt's House	Home ↔ Downtown
Distances involved:	.5 km each way = 1 km	1.5 km each way = 3 km	4 km each way = 8 km	5 km each way = 10 km
No. trips : per week	5	2	2	1
Total:	5 km	6 km	16 km	10 km
Mode used:	Bike	Walking	Bike	Bus/subway

Figure 10
Fill in an organizer like this one with the modes of transportation you use and the distances you travel in an average week.

Tourist Movement

Until the last 70 years or so, there were few **tourists**. Only those who were very wealthy and had lots of time to spare could travel great distances just to visit a special place. Today, tourism is the fastest-growing industry in the world. In 1989, the number of people who visited another country was 426 million. By 1998, this number had grown to 625 million. This increase in tourism is also occurring within country borders.

This growth is occurring because travel is much easier than it used to be. Many people have more leisure time than before, as well as more disposable income. There is also a growing awareness of the unique qualities of different parts of the world, which make people want to travel.

Tourist travel to another country ("international tourism") depends on airports that are **travel hubs**. These airports offer frequent flights to a large number of international destinations. Lester B. Pearson Airport is one of these travel hubs and is the busiest airport in Canada. O'Hare Airport in Chicago and New York's Kennedy Airport are important travel hubs in the United States.

Most people plan their vacations around when the weather is most pleasant in their travel destinations. Families with students usually must limit their travel times to school vacation periods. For both these reasons, the summer period is the busiest time for travel. Winter and spring are the next most popular vacation times. Airlines and hotels charge more during these *peak* times. For example, if you had taken a two-week vacation to Paris, France starting on July 1, 1999, Air Canada would have charged you return airfare of $2043.88 (from Toronto). The return airfare for the same trip starting on February 1, 2000 would have cost $981.88.

tourist—a person who visits another place. The visit lasts more than one night but less than one year.

travel hub—an airport that is at the centre of routes from many directions.

Figure 11
France is the world's top tourism destination.

Discover For Yourself

1. Why do you think travel hubs are in densely populated areas?
2. Why do you think airfares vary at different times of the year? What are the advantages and disadvantages to airlines and tourists?
3. In small groups, choose one of the top tourism destinations in Figure 12 and research it. Make a web showing at least five reasons why it is popular.

LANGUAGE LINK

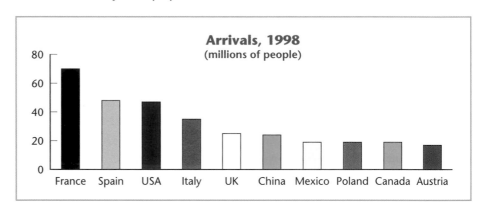

Figure 12
The World's Top Tourism Destinations

Summary

In this chapter you have learned about how modes of transportation vary around the world. You have seen the patterns of traffic in and around a large city. You have also discovered the importance of the growing tourism industry.

Reviewing Your Discoveries

1. In small groups, make a list of the modes of transportation mentioned in this chapter. Make an organizer that describes worldwide changes in the use of each mode of transportation between now and the year 2050. Include in the organizer explanations for why the changes are predicted to happen.
2. How do limited-access highways make commuting travel easier and faster?
3. Why has tourism become the world's fastest-growing industry?

Using Your Discoveries

1. Imagine you are an entrepreneur starting a travel business. You want to organize a vacation trip for 11 people, including yourself. You will do all the planning, and charge the 10 other people on the trip 10 per cent more than the trip actually costs. In that way, your own expenses for the trip will be covered.

   a) Choose three possible destinations for your trip: one in Canada, one in the United States, and one in another country. Use an organizer like the one in Figure 13 to help you decide which place is the best destination. Add two questions of your own to the first column to ensure you make a careful decision. You will have to do some research on the destination to answer the questions.

   b) Make your decision and explain in a written paragraph why the destination you chose was the best one.

   c) Make a newspaper advertisement or television commercial for the trip. It should include
   – details of the trip: destination, dates, cost, highlights
   – contact information for people to find out more

   d) Make a pamphlet that provides more details of the trip. In addition to the information in the newspaper or on television, it should include
   – the schedule and planned activities
   – arrangements for local travel at the destination

Factors decision will be based on:	Canadian Destination	American Destination	Other Destination
Is there enough interest in this destination and its features?			
Will the weather probably be good for our needs?			
Will I be able to supply any special equipment, or can I buy or rent it?			
Will there be 10 people interested with enough money?			

Figure 13
Add two questions of your own to this chart, then fill it in.

Chapter 6

Movement Milestones

Key terms

mobility

thrust

aerodynamic

In this chapter we focus on technological advances in travel. The information and activities will help you

▸ describe the development of travel
▸ describe how technology has improved mobility
▸ predict how mobility and travel might change in the future.

Moving Toward the 20th Century

Imagine taking a trip through time, looking out for changes in people's travel. You begin your journey over 5000 years ago, with a bird's-eye view of people moving over land and sea. You watch humans in different groups trading handmade goods for food supplies. They are trying to improve transportation to advance their trading possibilities. Their fastest mode of travel is using poles or oars to move boats, canoes, and rafts along rivers.

With the invention of the wheel 5000 years ago, people are transporting themselves and their goods in wagons pulled by oxen. These travel at a speed of about 3 km/h. In 3000 years' time, much faster horse-drawn chariots are being used. Over very long distances, donkey or camel caravans are travelling along trading routes. Meanwhile, on the sea, trading ships, powered by sails and oars, are becoming ever larger.

A little more than 2000 years ago, you see paved streets and road systems taking shape. Thousands of kilometres of roads are built throughout China, the Roman Empire in Europe and the Mayan Empire in Central and South America. But water trans-

Figure 1
About 500 years ago, Leonardo da Vinci filled notebooks with sketches of flying machines and submarines. But more than four centuries passed before these imaginings became reality.

portation remains a much faster and safer way of travelling. About 1500 years ago, paddle-wheel boats powered by animals are seen.

Over 200 years ago, you see a true transportation milestone: the steam engine. Steam engines make train travel possible. Thousands of kilometres of track are built during the 19th century, including railway bridges and tunnels. Within large cities, electric trams are used from the 1880s onward. The first car is seen in 1893. The subway is introduced in Paris in 1900, although underground railways have been operating already in London, England for some years. Ships continue to get bigger and faster.

 **For Yourself**

1. Make a timeline (using illustrations and text) to show the development of transportation from the beginning of modern human existence to 1900.

2. Figure 2 shows the development of bicycle technology in the 19th century.

   a) The first bicycle, the "hobby horse," had no pedals. How do you think people made it go?

   b) Pedals on Macmillan's bicycle were attached to the back wheel by a lever. In contrast, pedals on the "penny farthing" were attached to the front wheel. How else was the design of the penny farthing different? Why do you think this design change was made?

   c) Describe how the design of the "safety bicycle," very similar to today's bicycles, is different from the penny farthing. What do you think the advantages of this design are?

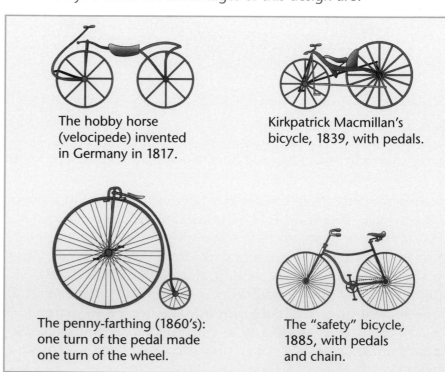

The hobby horse (velocipede) invented in Germany in 1817.

Kirkpatrick Macmillan's bicycle, 1839, with pedals.

The penny-farthing (1860's): one turn of the pedal made one turn of the wheel.

The "safety" bicycle, 1885, with pedals and chain.

Figure 2
The Development of Bicycle Technology in the 19th Century

Case Study　　*Travelling Back in Time*

Around the turn of the century, many European emigrants travelled to Canada. What modes of transportation did they use and how long did they take?

In 1897, the Basarabi family heard that there were large farms being given away in the Prairie provinces of Canada. They decided to leave their Romanian village southwest of Bucharest (Bucureşti). Their destination was Neepawa, a small settlement 180 km west-northwest of Winnipeg. The map in Figure 3 shows their route.

On their first day, they travelled 60 km by horse and wagon to Bucharest. The road was bumpy and dusty, and they were exhausted when they finally arrived that evening. The next day they took a train from Bucharest to L'viv in Ukraine. At the border, they were examined for diseases. This train ride took two days.

At L'viv, they switched trains to get to Krakow, Poland. This was another two-day trip, with another check for diseases at the border. One more train trip to go!

The train from Krakow was packed for the whole two days it took to get to Hamburg, Germany. After one more health check, the family boarded a ship that would take them across the Atlantic Ocean to Canada. It seemed like the whole world joined them as crowds jammed into the ship.

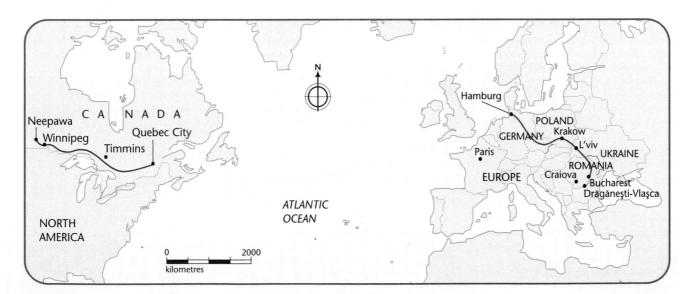

Figure 3
It took the Basarabi family 24 days to travel from their Romanian village to Neepawa, Manitoba.

The family's main memories of the ocean voyage were of rough seas, crowded conditions, poor food, and lots of sick people. The voyage took 13 days. When the ship docked at Quebec City, everyone had another medical examination. Then they took another crowded train trip to Winnipeg. For four days, the family rode with their fellow travellers, most of whom were from the United Kingdom. When they arrived at the train station, they went to an office nearby and applied for a farmstead near Neepawa. They hired a wagon to load up with the supplies they needed and rode to a hotel for the night.

The next day, they set off by wagon to Neepawa. The 200-km journey took two days as they wound their way over a dusty plain. Until roads were built, the Basarabi family and other homesteaders would be quite isolated on the Prairies. Would they be able to make a farm from the grassland?

Discover For Yourself

1. Copy the organizer in Figure 4 and fill in the blanks.

Day	From	To	Method of Travel	Comments
1	Drăgăneşti-Vlaşca		Horse and wagon	The road was bumpy and dusty. The journey took the whole day.
2–3		L'viv, Ukraine		
		Krakow, Poland		
		Hamburg, Germany		
				Rough seas. Very crowded, poor food, many sick. Medical examination in Quebec.
19–22				Crowded, many very poor people, mostly from UK.
		Hotel for the night	Wagon	Hired wagon to use to help to get supplies.
				Hot, bumpy, dusty, no road. Here at last!

Figure 4
Use the map on page 249 to fill in this chart.

2. Imagine that you are a travel agent in Romania today. You have to make travel arrangements for members of a Romanian family that have been accepted as landed immigrants to Canada. They live in Craiova and must get to Timmins, Ontario. These places are marked in Figure 3. Assume that the only link between

Craiova and Bucharest is a road—there is no passenger rail service to Bucharest. Also assume that air passengers from Bucharest to Canada must change planes in Paris.

a) Working in small groups, make an organizer like the one in Figure 4. Fill in the second column together as you figure out each stage in the travel arrangements for the family. Use a road map or other source of information to figure out the best way to get from Pearson International Airport to Timmins.

b) Each group member should choose one stage in the journey to research. For your stage, find out how long it takes. You may need to contact travel agencies and airlines or use road maps. (For the land stage(s), assume that these people would not be confident enough to drive very far.) In the "Comments" column of the organizer, describe the conditions that the travellers should expect at that stage of the journey.

c) As a group, draw a map that shows the route you have chosen. Use different colours to show different modes of transportation. Use Figure 3 on page 249 as an example. Include a legend, title, and north sign.

d) Discuss the ways in which the modern-day journey is different from the 19th-century one. Your discussion should include the changes that have taken place in transportation technology over the past 100 years. Summarize your discussion in one or two written paragraphs.

Milestones in Modern Transportation

There are two ways to improve **mobility**. One is to improve the mode of transportation. The other is to lessen the barriers that prevent or slow down movement.

The 20th century's major improvement to modes of transportation was air and space travel. Air flight became possible with the invention of mechanical wings that are curved on top and flat beneath. The invention of the jet engine, which uses **thrust**, made air flight much faster. Speed was increased even more when airplanes were given an **aerodynamic** design.

In the Concorde's aerodynamic design, the swept-back, triangular-shaped wings give a better airflow for high-speed flight.

LANGUAGE LINK

mobility—the ability to move.

thrust—a force created in a jet engine that drives or pushes a vehicle forward.

aerodynamic—designed to make the movement of a vehicle through air fast and smooth.

Figure 5
A Concorde Jet

The Concorde can travel at twice the speed of sound, crossing the Atlantic Ocean in three hours. As there is a five-hour time difference between London and New York, a Concorde taking off from London at 10 a.m. arrives in New York two hours "earlier," at 8 a.m.

Improvements to modes of transportation would have been of little use without the construction of roads, railways, ports, canals and airports.

Canals have played a vital part in improving water routes around the world. The Suez and Panama Canals cut thousands of kilometres and hours from sea voyages. Canals with locks to control water flow have enabled ships to travel in places that were previously too shallow or where there was no existing water route.

Ship travel from the Atlantic Ocean to the head of Lake Superior was greatly improved by the opening of the Great Lakes-St. Lawrence Seaway in 1959. Through the use of canals, locks and control dams, the seaway provides a deep waterway that ships can navigate to transport raw materials and finished goods.

Tunnels and bridges create new transportation routes between places separated by mountains or water. The latest major achievement in tunnel technology was the English Channel Tunnel (1994). One of the most recent achievements in long bridges was the 13-kilometre-long Confederation Bridge between New Brunswick and Prince Edward Island (1997). Before the bridge was opened, the journey by ferry to Prince Edward Island took 45 minutes and often involved hours of delays. Bad weather made the delays longer still. The road journey now takes 12 minutes, and there are infrequent delays.

The Future of Transportation

The future of transportation is being tested today. For many years, scientists and engineers have been trying to make a car that does not pollute. One solution that has had some success involves the use of electricity to power cars. The car is plugged into an electrical outlet when not in use and runs on battery power. On the downside, electrically powered vehicles now on the road cannot go as quickly as regular vehicles. They also have to be charged up frequently, and the production of the electricity they need can cause pollution.

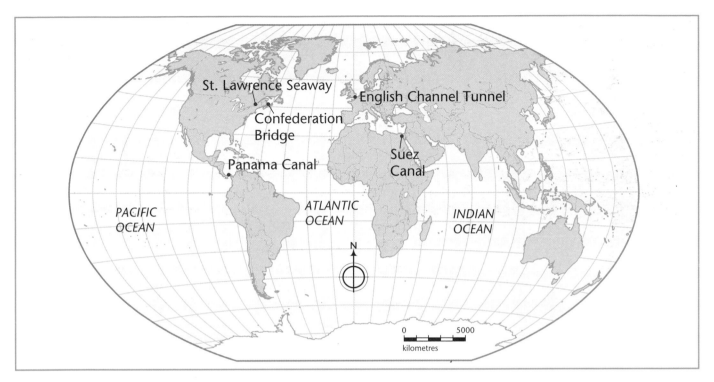

Figure 6
Describe how each of the technological achievements marked on the map overcame a barrier to movement.

Solar-powered cars are an option in sunny climates. But they have the same battery problems as electric cars—not much power can be stored in the battery, so recharging is frequently needed.

During your lifetime, space travel may become a reality for more people than ever before. With the advent of the re-usable space shuttle, the cost of space travel was reduced. The permanent space station may prove to be another crucial step in human exploration of the solar system.

While a few people will have the opportunity to travel to other worlds, most of us will try to improve our mobility on Earth. More vehicles will result in more traffic and more pollution, unless we can find new technologies to deal with these problems. Developing such technologies requires people who can think in new ways. Many ideas will fail, but some will succeed. Upon these successes our future mobility depends.

Discover ☀ *With Photographs*

1. Figures 7 and 8 show two transportation ideas that did not become milestones. In small groups, discuss these questions about them.

   a) Why do you think moving sidewalks are not used in cities? Include ideas about expenses and climate in your answer.

   b) Where are moving sidewalks used? Why do you think they became part of these environments?

   c) Rocket and jet belts were developed from the 1950s through the 1970s. Why do you think they are not used today? Include ideas about their purpose, fuel, and landing in your answer.

   d) Based on your answers to (a)-(c), make a list of criteria that a new transportation idea would have to meet in order to catch on.

2. In your groups, use the criteria you listed in Question 1 (d) to help you to think of three new transportation ideas. For each one,

   a) describe the idea. Include in your description a specific traffic or travel problem that the idea would solve. Also include a sketch of the idea.

   b) describe the effect the idea would have on travel in the future.

Figure 7
What are the advantages and disadvantages of moving sidewalks?

Summary

In this chapter you have discovered how mobility has changed as a result of technological advances. You have learned about some achievements and failures in our ongoing pursuit of improved mobility. You have also had a glimpse of what may be around the corner in the world of transportation.

Reviewing Your Discoveries

1. Give an example of how technology has improved a vehicle that is used for moving people.

Figure 8
What are the advantages and disadvantages of rocket belts?

2. Give three examples of different routes that have been created or improved to allow vehicles to travel more easily.
3. List two problems that scientists and engineers must keep in mind in designing future transportation.

Using Your Discoveries

1. Science fiction stories often deal with how people will move in the future. Two mobility ideas inspired by science fiction stories are summarized below. You may want to find other stories with mobility ideas and summarize them for this exercise. For each mobility idea,
   a) Write one paragraph about whether or not you think people will welcome the idea described. Give reasons for your opinion.
   b) Write a second paragraph about the advantages and disadvantages of the idea.

Idea #1
In the future, people never have to travel to go on vacation. Instead, they go to a "virtual vacation" agency that gives them memory implants of having been to a particular travel destination. The implants give them the kinds of memories they might have had if they had actually travelled to the destination.

Figure 9
Do you think this travel idea will catch on?

Idea #2
In the future, scientists have learned to combine genes from bird species with human genes. With careful genetic engineering, certain humans have wings and can fly like birds.

Figure 10
What would be the advantages and disadvantages of this idea?

Geography Workshop

This unit has presented information on many aspects of human movement, including these five:

▶ the movement of cultural features across the world, resulting in multicultural societies

▶ the movement of international refugees and migrants within countries because of war and famine

▶ the movement of immigrants and temporary migrants to find work and other opportunities

▶ the modes of transportation used in human movement

▶ the effects of technology on modes of transportation and movement routes.

In this **culminating activity**, you will use your knowledge of human movement to describe and explain the experiences of a specific cultural group in Canada.

a)

Estonians in Canada

Date: June 1

Names: Tiina Randoja, Jenny Amin, Stewart Coe

Cover page introduces the scrapbook with a title, date, and your name. It should be decorated to reflect the contents of the scrapbook.

d)

Special Characteristics and Tra[...]

For a long time, Estonia was a rural or agri[...] celebrations and food traditions show how [...] Here are some examples of traditional Esto[...] recipe on the right.

Everyday Foods	Holiday Foods
Boiled potatoes	Blood buns
Milk soup	Blood sausage
Pearl barley porridge	Brawn
Rye bread	Roast pork or roast goose
Salt herring	Sauerkuart
Wild mushrooms	Apples, mandarin oranges
Yoghurt	Gingerbread, nuts

Traditional costumes are worn on festive o[...] shows people wearing costumes at the Nati[...] every five years in the summer.

c)

Reasons For Coming To Canada

The reasons why Estonians have come to Canada have depended on the time period in which they came.

1945 to 1954

After the end of World War II, there were over 70 000 Estonian refugees, mostly in Germany and Sweden. Most of them lived in "displaced persons camps," where life was hard, and the refugees were worried about the future. They did not want to be sent back to Estonia because it was now part of the Soviet Union under Stalin. They thought life in Canada would be safer than life in the camps.

Timeline of Important Events

1945	World War II ends
1946	A group of Baltic nationals in Sweden are forced to return to the Soviet Union.
1945-46	Stream of immigrants first arrives in Nova Scotia in "Viking boats."
1991	Estonia becomes an independent nation.

The first immigrants to Canada after World War II came in small boats. Some were only 9 to 20 metres long.

The Canadian government helped pull Estonians to immigrate by paying for them to move to Canada in exchange for two years of work.

continued on next page

Illustrations and text show the **reasons** that members of the group decided to leave their homeland and come to Canada and any **notable experiences** they had on the way.

Maps, illustrations, and text show the **place of origin** of the group, the **years of their migration** to Canada, the **routes** they took and the **modes of transportation** they used.

Illustrations and text show the **special characteristics and traditions** of the group in their original area.

Illustrations and text show how and why the **culture has changed** since the group arrived in Canada and the **contributions to Canada** the group has made and continues to make.

b)

Place of Origin

Estonians come from Estonia, a country in northeastern Europe. The mainland area is about 45 100 square kilometres. There are also about 800 islands in the Baltic Sea near the mainland. It is a land of many lakes and rivers, with a continental climate, like much of Canada. Today Estonia is an independent country, but from 1944 to 1991 it was a republic of the Soviet Union, called the E.S.S.R. (Estonian Soviet Socialist Republic). During those years, Estonians were not allowed to emigrate.

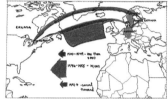

Migration to Canada

Between 1900 and 1944, less than 3000 Estonians immigrated to Canada.

During World War II, 72 000 Estonians escaped their homeland. They were political refugees and "displaced persons" who at first settled in Sweden and Germany.

continued on next page

e)

Contributions to Canada

Compared to other immigrant groups, Estonian immigrants to Canada are among the most highly educated. They have made contributions to academic study, amateur sports, architecture and other areas. Here are some examples.

Rhythmic Gymnastics

Rhythmic gymnastics began in Eastern Europe about 100 years ago. It was developed by the Estonian Ernst Idla, who invented rhythmic dancing and founded the Institute of Gymnastics in Tallin, Estonia in 1929. One of his students, Evelyn Koop, immigrated to Canada and founded "Sport Club Kalev" in 1951. The club's first members were mostly Estonian girls, and the club was the only place in Canada where rhythmic gymnastics was practised. Now, Canadians of many backgrounds belong to rhythmic gymnastics clubs all over the country. Canada has been competing at World Championships in rhythmic gymnastics since 1971. Canadian Lori Fung, trained by Evelyn Koop at one of her clubs, received the first Olympic gold medal ever in rhythmic gymnastics in 1984. In 1996, another student of Evelyn Koop's, Camille Martens, was the only Canadian rhythmic gymnast at the Olympic Games.

Biography

Evelyn Koop

Evelyn Koop is famous for developing rhythmic gymnastics as a national sport in Canada. After founding Sport Club Kalev in 1951, she founded the Canadian Rhythmic Sportive Gymnastics Federation in 1968. In 1976, she directed and choreographed the opening ceremonies at the 1976 Montreal Olympics. This helped make rhythmic gymnastics an Olympic event in 1984. Koop's awards include the Queen Elizabeth Medal for Achievement, the Air Canada Sports Executive of the Year award and the Ontario Provincial Achievement Award.

continued on next page

Your Job

With one or two classmates, choose a cultural group (or a small part of a group) that has members or descendants in Canada. It could be one of the groups of which you or your ancestors are members. Create a scrapbook about the group that includes maps, illustrations, short pieces of writing and any other suitable items. The information in your scrapbook should include the group's

▸ place of origin
▸ special characteristics and traditions
▸ reasons for coming to Canada
▸ migration information: years, routes, modes of transportation, experiences
▸ ways in which the culture of the group has changed since arriving in Canada
▸ contributions the group has made and continues to make to Canada.

Staple or bind your finished scrapbook together, and include an attractive cover page. Put together an exhibit in which your group scrapbooks are on display.

Geography Skills

Review and Practice

Contours on Topographic Maps

Topographic maps allow us to examine an area of land and its man-made features, such as settlements, roads and bridges, etc. as well as its natural features such as hills, cliffs, and rivers. Many of these features will appear on the map as you would expect them to if you were looking at the area through the window of an airplane. For example, houses, barns and other buildings will appear as little black squares, lakes will be blue areas and roads will be generally straight or gently curving lines. However, natural features such as hills may not be as obvious on the map. That is because land height in the topographic maps is sometimes shown using contours.

Contours are lines joining places of equal height above sea level, measured in metres (or feet). Numbers beside dots or in small triangles give additional information about elevations.

Step-by-Step: Interpreting Contours

1. Examine the map and locate the contour lines. Each line will have a height measurement assigned to it, either in metres or in feet above sea level.
2. Note that land on one side of a contour is above the contour height, and below it on the other side.
3. If you can, find instances where the contour lines are drawn very close together. These lines indicate the land is sloping. The closer together the contours are, the steeper the slope is.
4. If the contour forms a "V" shape pointing to higher ground, the feature is a valley (well illustrated in Figure 3 on page 9).

5. If the contour forms a "V" shape pointing to lower ground, the feature is a high ridge of land such as a mountain or hill (for example, Feather Point at the end of the peninsula south of Bristol's Hope Cove in Figure 4 on page 10).
6. Contours never cross, but if there are places on the map where they join, this indicates a cliff.
7. Flat land usually has few contours, which tend to wander across the map (e.g. the 825 foot contour which passes east of Plum Coulee in Figure 7 on page 14).

Try This

1. Look at the narrow peninsula south of Harbour Grace in Figure 4 (on page 10). Which side of the peninsula has the steeper slope?
2. In Figure 3 (on page 9) there are a number of creeks running into Lake Erie. What problems might these creeks cause for farming in the area?
3. There is an artificial drain running across Figure 7 (on page 14). Why do you think this is necessary?

Satellite Images

Many satellites exist above the Earth today. Some are communications satellites, which usually remain in one place to transmit telephone signals or TV pictures from one country to another. Others, like the US *Landsat* or the French *Spot* satellites, orbit the Earth to scan the features on the Earth's surface. They send digital data that can be displayed as either true colour or false colour images. These images can provide much information about the earth's surface, including rocks, vegetation cover, air and water pollution, as well as human features such as cities.

The *Landsat* satellite scans the surface of the Earth in 30 m^2 sections. The *Spot* satellite can scan an area as small as 10 m^2.

Figure 1
Satellite Image of Ottawa

Try This

1. On the satellite image in Figure 1,
    a) Identify the central business district, Parliament Hill, the Rideau River, and Rideau Canal.
    b) What transportation feature is located in the extreme northeast of the image?
    c) Compare this image with your map of population density in Ottawa (Figure 7 on page 58). How do high population density and low population density areas differ on the image?
    d) How does the road pattern in newer suburbs differ from the pattern in older residential areas?
    e) Compare this image with the one of Winnipeg on page 42. Which image is true colour and which is false colour? Why might false colour images be used?

Circle Graphs

A circle graph consists of a circle that is divided up into sections which resemble the slices of a pie. The size of each slice shows how much that item makes up as a part of all of the items being graphed. For example, the circle graph in Figure 2 clearly shows the comparative areas of the provinces and territories that make up Canada.

Actual numerical values, such as area (in km^2) or percentages, may be added to the graph to make it possible to read information from it accurately. In this example, the percentage values have been included. The area (in km^2) could have been written beside each sector.

Step-By-Step: Drawing a Circle Graph

1. You may be given data that has already been converted to percentages. If not, you will have to do this yourself by dividing each item by the total value of all the items, then multiply the result by 100. For example, to convert the land area of Ontario to a percentage, you would divide the area of Ontario (in km²) by the total area of Canada (km²) and then multiply the result by 100.

2. Once you have the data in the proper form, you can begin plotting it on the circle. Use a ruler and pencil to draw a line from the centre point to the top of the circle.

3. On a circle that has 100 divisions marked around the perimeter, count clockwise around the edge of the circle and make a small mark at the first of the values that you must plot. (If you do not have a circle with 100 divisions marked on it, you must multiply each percentage value by 3.6 to calculate the angle that you will plot using a protractor.)

4. Use a pencil and ruler to draw a line from the centre to this point on the circumference of the circle.

5. Count from this new line until you reach the next value. Make a mark and draw a line from the centre as you did previously.

6. Repeat step 3 until all the data are plotted. If you are correct, the last line will be in the same place as the first line.

7. Finish your graph by colouring and labelling each section. Add any data as instructed and include a suitable title.

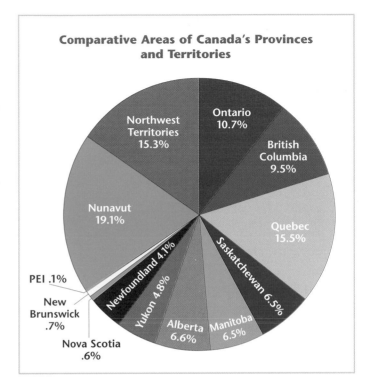

Comparative Areas of Canada's Provinces and Territories

Figure 2
A Circle Graph Showing the Comparative Areas of Canada's Provinces and Territories

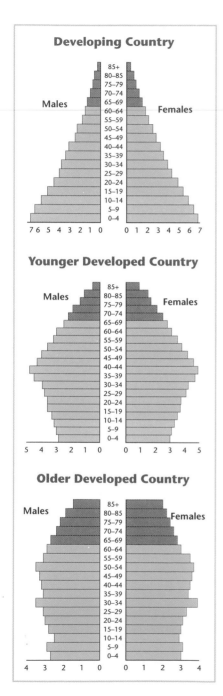

Figure 3
Model Population Pyramids

Population Pyramids

Population pyramids are a graphic way of showing how a population is divided up by sexes and age groups. They are sometimes referred to as "age-sex pyramids."

Pyramids fall into three of categories according to their shapes (see Figure 3 on this page). First, developing countries tend to have a genuine pyramid shape, like the one for Mozambique on page 73. Second, younger developed countries, such as Canada, have pyramids that are narrower toward the base, with a bulge for the "baby boom" generation (who were in their 30's and 40's by 1996). Finally, an older developed country, such as Sweden, has a greater percentage of older people, giving the pyramid a "top heavy" shape.

Step-By-Step: Drawing a Population Pyramid

1. If need be, convert your data to percentages of the total population. For example, in Figure 3 (on page 74), males over 85 as a percentage of total population has been calculated as follows:
   109.5 (males over 85) ÷ 29969.2 (total population of Canada) × 100 = 0.4
2. Draw a horizontal axis on graph paper in which 1 cm represents 1% of the age/sex group. The scale begins at 0 at the centre bottom of the graph paper, and counts to the left for males and to the right for females, as on Figure 2 (page 73).
3. Draw a vertical axis upward from the zero point on the horizontal axis. You may wish to draw two vertical axes, leaving space for writing in the age groups in between, as in Figure 3 on the left. The scale on the vertical axis should be 0.5 cm for each 5-year age group.
4. Beginning with the 0–4 age group, draw horizontal bars, with males to the left and females to the right of the vertical axis.
5. Continue up through the age groups.
6. You may wish to shade in your graph, either by males and females, or by age groups such as 0–14, 15–64 and 65+ (as in Figure 3 on this page).

Try This

1. Which of the three shapes in Figure 3 do you think population pyramids for each of the following countries would have? Give reasons for you answers.
   a) United States
   b) Zambia
   c) England
   d) Japan
   e) Mexico
2. Work out how the following events would affect a population pyramid. Draw a rough sketch of what you think the pyramid might look like 20 years after each event.
   a) World War One and its effects on the population of France
   b) The policy, initiated in China in 1979, which allowed each couple to have only one child
   c) The Famine in Ethiopia in 1984–85
   d) The AIDS epidemic presently affecting a central African country.
   You can work in groups to decide on the shape of your sketches. Give reasons to justify the pyramids you have drawn.

Scattergrams

A scattergram is a type of graph that helps us to judge whether or not there is any relationship between two factors, such as latitude and midsummer temperature or the area of a country and the size of its population. A scattergram consists of dots, each representing one piece of data. If the dots seem to show a connection between the two sets of factors, a line can be drawn through the plotted data to show that trend.

Step-By-Step: Plotting a Scattergram

1. Draw two axes at a suitable scale, one to represent each factor.
2. Plot dots to represent the information that you have been given.

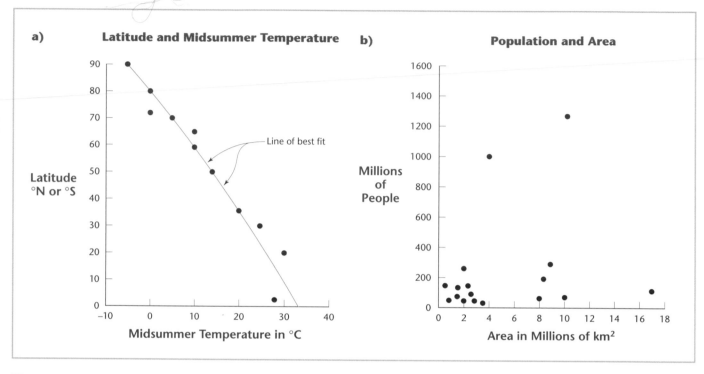

Figure 4
The scattergram in (a) shows there is a close relationship between the two factors: As latitude increases, temperatures decrease. In (b) there is no obvious relationship between the size of a country and its population.

3. If you observe that there is a trend evident (the majority of dots seem to increase or decrease in the same general direction), draw a line which may be straight or gently curved through the mid-point of the dots.

Compound Bar Graphs

A compound bar graph consists of vertical or horizontal bars each of which is sub-divided into sections rather like a stack or line of bricks. Each section shows how much of the total value is made up of that item.

Data for compound bar graphs may be in the form of the original figures as in Figure 5a or it may be converted to percentages, as in 5b. If each bar shows change through a time sequence, dotted lines are often used to connect the "bricks" in the bars as is shown in Figure 5c.

Step-By-Step: Plotting a Compound Bar Graph

1. Set up the two axes of your graph so that there will be room to plot the total values of each bar.
2. Plot the total value of the information to make the first bar.
3. Within the bar that you have already plotted, draw a straight line across at the value of the first item that makes up a part of the total value for that bar.
4. *Starting at the line that you plotted in the previous step*, count up the value of the next item that makes up a part of the total value for that bar. Alternatively, you could add these last two values to find out where to plot this line.
5. Repeat step 4 until all values have been plotted. If you are correct, your last line will be the same as the top of the bar.
6. When you have finished plotting all the bars, colour the "bricks" using the same colour for the same item in each bar.
7. If the bars show a time sequence, you may use dotted lines between bars to emphasize the changes that are shown.
8. Add labels and a suitable title.

Figure 5
Three types of Compound Bar Graphs

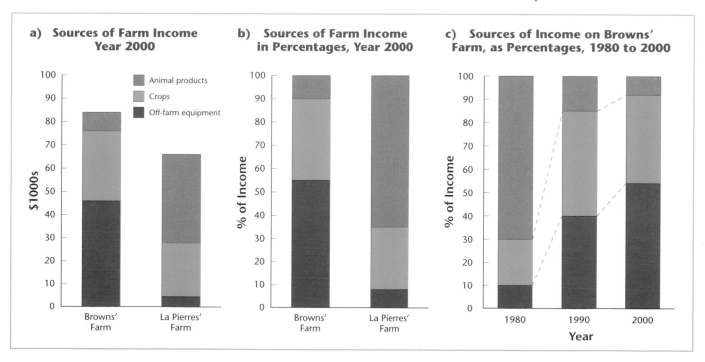

a) **Sources of Farm Income Year 2000**

- Animal products
- Crops
- Off-farm equipment

b) **Sources of Farm Income in Percentages, Year 2000**

c) **Sources of Income on Browns' Farm, as Percentages, 1980 to 2000**

Try This

1. Look at the three examples of compound bar graphs in Figure 5 and answer these questions. Which graph: a, b, or c would be best for
    a) showing how the proportions of different immigrant groups in Canada has changed since 1900?
    b) showing the different proportions of immigrants from different cultures in selected Canadian cities in the year 2000?
    c) showing the numbers of immigrants from different cultural groups in each province and territory?
2. Sketch a compound bar graph you could use to illustrate how the number of immigrants from different cultures has changed in Canada since 1950.

Glossary

aerodynamic designed to make the movement of a vehicle through air fast and smooth.

baby boom a period of about 20 years, following World War Two, during which there was an unusually high number of births. It was followed by the "baby bust"—a period of significantly fewer births.

bilingual involving two languages. In Canada, this term refers to French and English.

birth rate the total number of births per thousand people in a country's population. The formula for calculating it is **Total Births ÷ Total Population x 1000**.

census a count of all the people in a country, together with some details about them.

central business district (CBD) the central part of a city that is the focus of transport routes. It contains the city's most valuable land and many of its leading businesses.

civil war armed conflict between people within a nation. It may be started by a group that wants to take power from the current rulers or by a group that wants another group to leave the country.

clustered settlement a pattern of settlement in which houses and other buildings are laid out closely together.

collateral something owned by a person taking out a loan that can be seized if the person does not repay the loan.

command economy an economic system in which the government owns and controls all parts of the economy.

commuting the daily movement to or from a place of work or study.

correlation a relationship or pattern between two factors that is fairly predictable. The factors compared in correlations are usually described by sets of numbers.

culture the behaviour that people learn, made up of their belief systems, languages, social patterns, political systems, organizations, food and clothing customs, and use of buildings, tools, and machines.

death rate the total number of deaths per thousand people in a country's population. The formula for calculating it is **Total Deaths ÷ Total Population x 1000**.

diffusion the spread of a new feature from a centre or centres. For example, the use of chili peppers for cooking and eating began in Central America about 5000 years ago. Explorers and settlers caused it to spread (diffuse) to Europe. It then diffused to South Asia with Portuguese settlers, and is now a basic ingredient in much South Asian food.

discrimination treating a group or individual unfairly based on their background.

economics the study of the production, movement, distribution, marketing and consumption of goods and services. Specifically, economics deals with *which* goods and services we produce, *how* we produce them, and how we *distribute* them.

entrepreneur a person who takes a risk to start and run a business.

exports things sent out of a country.

foreign investment one country allowing people from other countries to *invest*, or put money into, its industries.

franchise the right to sell a particular good or service. Each franchise operation must follow guidelines to make it similar to other franchises owned by the same company.

gross national product (GNP) the sum of the value of all the goods and services produced in a country in a year. It is often measured in American dollars (US$).

immigration the act of people entering and settling into a country different from their native country.

impact benefits agreement a voluntary agreement signed by a mining company and a local Aboriginal group. The mining company agrees to minimize the mine's impact on ecosystems and may guarantee employment to local residents and funding for environmental research.

imports things brought into a country.

indigenous peoples cultural groups who lived in an area from early times before the arrival of colonists.

industrialized nations countries that use high levels of technology in all sectors of the economy.

Industrial Revolution a set of changes in technology, social life, and politics that occurred during the late 18th and early 19th centuries. During this time coal was used to power the steam engine and many other mechanical inventions. As a result, Europe, particularly Great Britain, became the leading industrial region of the world.

inputs the factors of production put into a manufacturing system. Inputs get worked on by processes.

kimberlite pipe a roughly cylindrical plug of a rare rock called "kimberlite." The pipe-shaped plug is formed by cooling magma and may contain diamonds.

landscape what we see when we look around.

life expectancy the average number of years a person is likely to live. It depends on many factors, particularly the standard of living in a person's country of residence.

linear settlement a pattern of settlement in which homes and other buildings follow the lines taken by roads.

literacy rate the percentage of adults (people over the age of 15) who can read and write.

market economy an economic system in which private individuals set up, own and direct businesses to produce goods and services that consumers want. This system is also called "free enterprise" or "capitalism."

migration movement from one area to another.

mixed economy an economic system that combines private ownership with government control.

mobility the ability to move.

modes of transportation the specific means by which people move from one place to another (for example, on foot, by bicycle, by car).

multicultural society a country or part of a country where large proportions of the population are from different cultural backgrounds. In such a society, people are encouraged or allowed to maintain their cultural traditions.

North American Free Trade Agreement (NAFTA) a trade agreement signed by Canada, the United States and Mexico.

official languages the languages in which the federal government conducts its business. Some provincial governments and many businesses use both official languages in their written material.

open-door immigration immigration that is free or unrestricted.

outputs the products leaving a manufacturing system that result from processes.

population density a measure of how many people live in a unit of area, usually a square kilometre.

population pyramid a set of two bar graphs placed back to back against a vertical axis. One shows the numbers of males, and the second shows the numbers of females, in different age groupings in a country.

primary industries industries that harvest raw materials or natural resources (e.g., agriculture, forestry, fishing, mining).

pull factors the social, political, economic, and environmental attractions of new areas that draw people away from their locations.

push factors the social, political, economic, and environmental forces that drive people away from one location to search for another one.

refugees people who have fled from their own country because of war, natural disaster or persecution based on race, religion, nationality, social group or political opinion.

ripple effect a chain of effects or events that results from an initial event. The chain is like the ripples caused by throwing a stone into a pond.

scarcity the limits in the amount of resources we have. While our resources are scarce (limited), our wants are great (unlimited).

scattered settlement a pattern of settlement in which houses and other buildings are placed a long distance apart from each other.

secondary industries industries that convert raw materials into finished products.

site the ground on which a building or city is built.

situation the location of a building or city in relation to surrounding places.

slave a person who is owned by another and who must do what the owner wishes.

spinoff a product or an idea resulting from an activity that was designed for a different purpose.

standard of living the extent to which people have the goods and services they need and want.

subsistence economy an economic system in which people's labour only produces enough, food, clothing, and shelter for the workers' own needs.

tariff a tax put on an import. By making imports more expensive, tariffs protect a country's industries and jobs.

tertiary industries industries that provide services (e.g., banking, retailing, education).

thrust a force created in a jet engine that drives or pushes a vehicle forward.

tourist a person who visits another place. The visit lasts more than one night but less than one year.

traditional economy an economic system in which people's methods of working have changed little from one generation to the next. Workers in a traditional economy try to produce a little more than what is needed for subsistence.

travel hub an airport that is at the centre of routes from many directions.

urbanization the growth of cities. This term can also mean the adoption of an urban lifestyle.

utilities things that are useful to us. City utilities include gas, water, electricity, cable, and telephone systems.

Illustration Credits

p. 13: Portion of map, Plate 26, 'The Railway Age, 1834–1891' from *Historical Atlas of Canada II* (Toronto: University of Toronto Press, 1993). Reprinted by permission of University of Toronto Press Inc.; p. 29: Map, 'Canada Population' adapted from *Canadian Oxford School Atlas*, 7th Edition, by Quentin Stanford (Toronto: Oxford University Press Canada, 1998).; p. 44: Adaptation of Figure 16, 'Urban and rural population, world, 1955–2015' from http://www.who.int/whr/1998, reprinted by permission of the World Health Organization.; p. 46: * Figure, 'Urban population (billions) 1950–2030' adapted from http://www.undp.org/popin/ wdtrends/ura/uracht2.htm.; p. 50: * Figure, 'World's urban agglomerations with populations of 10 million or more inhabitants in 1996: 1970, 1996 and 2015' adapted from http://www.undp.org/ popin/wdtrends/urb/ urbcht1.htm.; p. 54: Map, 'Winnipeg' from *Canadian Oxford School Atlas*, 7th Edition, by Quentin Stanford (Toronto: Oxford University Press Canada 1998).; p. 54: Map, 'Halifax' from *Canadian Oxford School Atlas*, 7th Edition, by Quentin Stanford (Toronto: Oxford University Press Canada, 1998).; p. 55: 'Map 1 (As Amended) Land Development Concept' courtesy of Plan Edmonton, Edmonton's Municipal Development Plan, City of Edmonton.; p. 73: Figure, 'Population Pyramid for Mozambique (1995)' adapted from *1997 UN Demographic Yearbook*, Table 7. The United Nations is the author of the original material. Reproduced by permission of the United Nations.; p. 82: Figure, 'World Population Growth, 1850–2050' adapted from *Population and Environment Dynamics* by Diana Cornelius and Jane Cover (Washington, DC: Population Reference Bureau, 1997). Reprinted by permission of Population Reference Bureau.; p. 88: Map, 'Economic well-being Indicator: Child Malnutrition: Prevalence of Underweight Under 5s', copyright OECD. Material available on OECD website at http://www.oecd.org/ dac/Indicators/ htm/map4.htm.; p. 90: * Figure, 'Number of persons chronically undernourished in developing countries' from http://www.fao.org/NEWS/FACTFILE/FF9609-e.htm.; p. 91: * Figure, 'One woman's day in Sierra Leone' adapted from http://www.fao.org/NEWS/FACTFILE/ FF9719-e.htm.; p. 92: Adaptation of Figure, 'Malaria Distribution and Reported Drug Resistance' from http://www.who.int/ctd/html/malariageo.html, reprinted by permission of the World Health Organization.; p. 94: * Figure, 'Generational impact of educating girls' adapted from *The State of the World's Children 1999* (New York: UNICEF, 1999).; p. 95: Figures, 'Net primary enrolment, by region (around 1995)' and 'Reaching grade five, by region (around 1995)' from United Nations Children's Fund, The State of the World's Children 1999, UNICEF, New York, 1998. Reprinted by permission.; p. 111: Figure, 'General Indicator GNP per Capita' from http://www.oecd.org/dac/Indicators/htm/ map_a.htm © OECD. Reprinted by permission.; p. 111: Figure, 'Share of Population, Resource Consumption, and Waste Production' adapted from *World Population and the Environment* (Washington, DC: Population Reference Bureau, 1994). Reprinted by permission of Population Reference Bureau and Natural Resources Defense Council.; p. 112: * Figure, 'Number of People per Vehicle, in Selected Countries, 1994' adapted from World Population and the Environment (Washington, DC: Population Reference Bureau, 1994, and Detroit, MI: American Automobile Manufacturers Association, 1994).; p. 116: * Map, 'The Commercial Empire of the St Lawrence' adapted from *A Historical Atlas of Canada* by D.G.G. Kerr (Scarborough, ON: Nelson Canada, 1961).; p. 122: Map, 'Canada Manufacturing' adapted from *Canadian Oxford School Atlas*, 7th Edition, by Quentin Stanford (Toronto: Oxford University Press Canada, 1998).; p. 130, 138: * Excerpts adapted from 'The Entrepreneur's Quiz' by William E. Jennings from http://www.strategis.ic.gc.ca/SSG/ mi03296e.html.; p. 135: Figure, 'Canadian Trade with Indonesia, 1993–1997' adapted from *Global Links: Connecting Canada* by Robert Kolpin. Copyright © Oxford University Press Canada 1999. Reprinted by permission.; p. 173: Figure, 'Mining at Ekati Diamond Mine' courtesy of BHP Diamonds, Inc.; p. 178: Excerpts from 'Brave New Brands' by John Schofield, *Maclean's*, 18 May 1998, p. 32. Reprinted by permission.; p. 185: Figure, 'Major Centres Where Different Cultures Developed' from *The Human World: A Changing Place*, 1st Edition, by Robert Harshman and Christine Hannell. © 1985. Reprinted with permission of Nelson Thomson Learning, a division of Thomson Learning. Fax 800–730–2215.; p. 186: Map, 'Religion' adapted from *Canadian Oxford School Atlas*, 7th Edition, by Quentin Stanford (Toronto: Oxford University Press Canada, 1998).; p. 186: Map, 'Language' adapted from *Canadian Oxford School Atlas*, 7th Edition, by Quentin Stanford (Toronto: Oxford University Press Canada 1998).; p. 195: * Map adapted from 'Nenets: Surviving on the Siberian Tundra' by Fen Montaigne from *National Geographic*, March 1998. Reprinted by permission from NGS CARTOGRAPHIC DIVISION/NGS Image Collection.; p. 203: Graphs from Chart 1.3, 'Inflows of migrants by country of origin to selected OECD countries, latest available year' from 'SOPEMI, Trends in International Migration' from *Annual Report 1998*, OECD, pp. 19 and 20. Copyright OECD 1998. Reprinted by permission of OECD.; p. 209: * Adaptation of 'Making Tracks: Migration in the 1990s' by Doug Stern from 'Human Migration' from National Geographic, October 1998. Reprinted by permission from DOUG STERN/NGS Image Collection.; p. 216: Figure, 'Criminal Offenses Against Foreigners in Germany Due to Prejudice, 1982 to 1996' reprinted by permission of European Forum for Migration Studies, University of Bamberg.; p. 218: Figure, 'Number of Foreign Citizens in Germany, 1982 to 1996' reprinted by permission of European Forum for Migration Studies, Bamberg University.; p. 223: 'Background/Portrait of Immigration' adapted from *The Globe and Mail*, 20 June 1992. Reprinted with permission from The Globe and Mail.; p. 226: * Maps, 'Most recent immigrants to Toronto (1991–1996)' from http://www.city.toronto.on.ca/ourcity/ profile2_maps.htm.; p. 228: 'They Come to Canada from Throughout the World' by Carrie Cockburn from *The Globe and Mail*, 24 May 1999. Reprinted with permission from The Globe and Mail.; p. 236: Figure, 'Greater Toronto Area' adapted from Appendix 1 from http://www.lbpia.toronto.on.ca/ Publications/Briefing%20Papers/market.htm.; p. 241: Illustration, 'World Traffic Volume...' by Laurel Rogers, from 'The Past and Future of Global Mobility' by Andreas Schafer and David Victor from *Scientific American*, October 1997, p. 61. Reprinted by permission of Laurel Rogers.; p. 248: Illustrations of bicycles from *Science and Technology: Black Holes to Holograms* (Oxford: Oxford University Press, 1993). Reprinted by permission of Oxford University Press.; p. 130: Excerpts adapted from *Entrepreneurship: A Primer for Canadians* by William E. Jennings (Toronto: Canadian Foundation for Economic Education, 1985). Reprinted by permission.

Photo Credits

Index

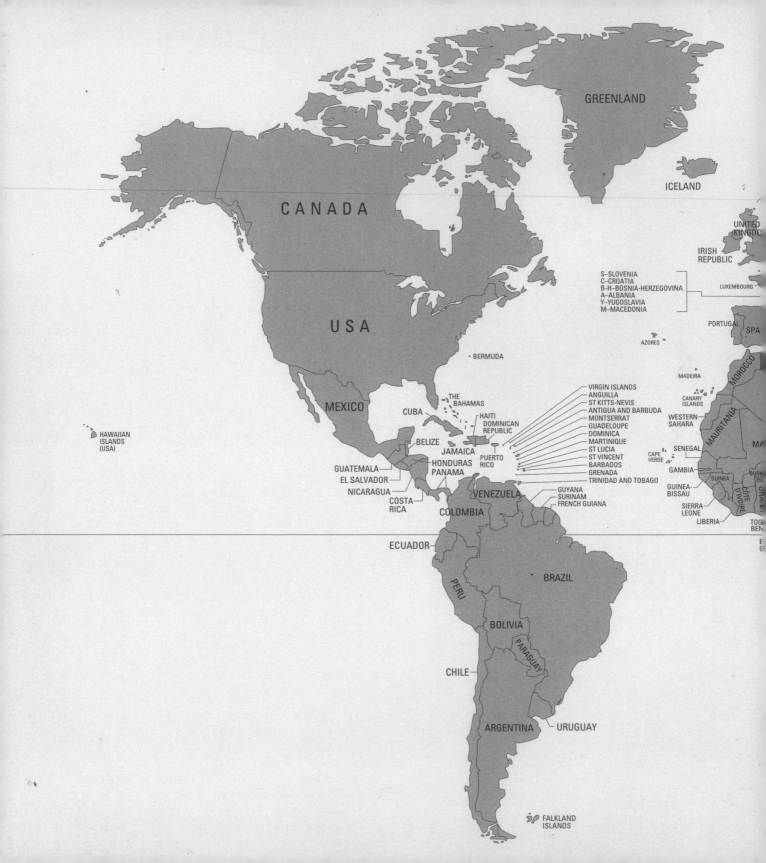

GREENLAND

ICELAND

CANADA

UNITED
KINGDOM

IRISH
REPUBLIC

LUXEMBOURG

S—SLOVENIA
C—CROATIA
B-H—BOSNIA-HERZEGOVINA
A—ALBANIA
Y—YUGOSLAVIA
M—MACEDONIA

USA

PORTUGAL

SPA

AZORES

• BERMUDA

MADEIRA

MOROCCO

MEXICO

THE
BAHAMAS

VIRGIN ISLANDS
ANGUILLA
ST KITTS-NEVIS
ANTIGUA AND BARBUDA
MONTSERRAT
GUADELOUPE
DOMINICA
MARTINIQUE
ST LUCIA
ST VINCENT
BARBADOS
GRENADA
TRINIDAD AND TOBAGO

CANARY
ISLANDS

WESTERN
SAHARA

CUBA

HAITI
DOMINICAN
REPUBLIC

MAURITANIA

MA

HAWAIIAN
ISLANDS
(USA)

BELIZE

JAMAICA

PUERTO
RICO

CAPE
VERDE

SENEGAL

GAMBIA

GUATEMALA
EL SALVADOR

HONDURAS
PANAMA

GUINEA-
BISSAU

GUINEA

BURK
FAS

NICARAGUA

COSTA
RICA

VENEZUELA

GUYANA
SURINAM
FRENCH GUIANA

SIERRA
LEONE

LIBERIA

CÔTE
D'IVOIRE

COLOMBIA

TOG
BEN

ECUADOR

B

PERU

BRAZIL

BOLIVIA

PARAGUAY

CHILE

ARGENTINA

URUGUAY

FALKLAND
ISLANDS